COVER PHOTO:
*With compliments of R.E. Turner
Vice Chairman Time Warner Inc.*

RIDING A WHITE HORSE

Ted Turner's GOODWILL GAMES & Other Crusades

Althea Carlson

IN HIS OWN WORDS

Riding a White Horse:
Ted Turner's Goodwill Games and Other Crusades
Copyright @ 1998 by Althea Carlson

First printing, May 1998

Cover design by Joy Allen, Immedia, Inc.
Cover photo by Mark Hill
Editorial Consultants: The Mackenzie Group
Published by: Episcopal Press
 222 Seacrest Drive
 Wrightsville Beach, NC 28480

Printed on recycled paper and bound in the United States of America. All rights reserved. No part of this book may be reproduced in any form or by any electronic or mechanical means including information storage and retrieval systems without permission in writing from the publisher, except by a reviewer, who may quote brief passages in a review.

DISCLAIMER: Any inaccuracies are not the intent of the author, but due to the sources used. All sources are annotated.

Library of Congress Cataloging-in-Publication Data

Carlson, Althea
 Riding a White Horse: Ted Turner's Goodwill Games and Other Crusades
by Althea Carlson
 272 pages, unillustrated
 ISBN 0-9663743-0-4
1. Turner, Ted. 2. Cable News Network – Biography
3. Businessman – United States – Biography
I. Turner, Ted. II. Carlson, Althea. III. Title.
Library of Congress Catalog Card Number 98-70400

PRE-PUBLICATION COMMENTS

"Who else but Ted Turner to capture the essence of Ted Turner? ... in all his quintessentially-American glory, between the covers of Althea Carlson's sparkling new book."

> Anne Russell, PhD
> Professor of Literature
> University of North Carolina

"Breathless ... a quick read."

> Peter Moon, president
> Sky Books, New York

"Most of the athletes (from the 1980 Moscow Olympics boycott) weren't bitter, but they had missed their chance. They had to get on with their lives."

> Floyd (Chunk) Simmons, decathlete
> Olympic decathlon bronze medalist 1948
> Olympic decathlon bronze medalist 1952
> North Carolina Hall of Fame

*Dedicated to
the 3500 athletes who competed
with bravaderie and sportsmanship
at the Moscow Goodwill Games
at the height of the US/Soviet Cold War
and to my own three athletes and sailors
Allyn, Bill and Stephen*

CONTENTS

AUTHOR'S NOTE
Page 1

INTRODUCTION
Page 7

CHAPTER ONE
PUNCTURING THE IRON CURTAIN WITH THE GOODWILL GAMES
Page 9

CHAPTER TWO
A BORN CRUSADER
Page 25

CHAPTER THREE
LIGHTNING STROKES, SNEAK ATTACKS
Page 39

CHAPTER FOUR
RADIO AND TV – TALKING BILLBOARDS
Page 51

CHAPTER FIVE
THE BRAVES, THE HAWKS, THE FLAMES, THE CHIEFS
Page 61

CHAPTER SIX
CAPTAIN COURAGEOUS – UH – OUTRAGEOUS
Page 77

CHAPTER SEVEN
TED TURNER DISCOVERS THE WORLD IS ROUND
Page 97

CHAPTER EIGHT
CNN CREATES A GLOBAL TALKATHON
Page 109

CHAPTER NINE
BOY ARE THOSE FOREIGN KINGS AND PRESIDENTS GOING TO BE SURPRISED TO SEE ME!
Page 129

CHAPTER TEN
SHAKING UP THE NETWORKS
Page 145

CHAPTER ELEVEN
THE CABLE NEWS WAR, CBS BID AND MGM
Page 153

CHAPTER TWELVE
TEN PLEDGES TO THE PLANET
Page 173

CHAPTER THIRTEEN
"SAVE THE EARTH" CAMPAIGN
Page 175

CHAPTER FOURTEEN
DREAM-O-VISION, TV AT GROUND ZERO
Page 181

CHAPTER FIFTEEN
HOME – AT LAST
Page 191

CONCLUSION
PRINCE OF THE GLOBAL VILLAGE
Page 205

AFTERWORD
Page 211

APPENDIX
1998 NEW YORK GOODWILL GAMES: LOCATIONS, DATES, EVENTS
Page 213

NOTES ON SOURCES FROM RESEARCH
Page 217

BIBLIOGRAPHY
Page 241

INDEX
Page 243

ABOUT THE AUTHOR
Back page

AUTHOR'S NOTE

One of the best ideas Ted Turner ever had never happened. He called together a group of script writers and film producers. Seated at a round table in his office he began, "All these missiles all over the globe, all the fear, it's terrible. It interferes with business, and it interferes with people's lives. What the world needs is – PEACE." He proposed making a movie in the genre of "Dinner With Andre," a long conversation with three characters; himself, Ronald Reagan and Mikhail Gorbachev. Turner would take film crews to Moscow and talk with Gorbachev about his problems with the U.S., asking why the Soviets are threatening nuclear attack. Then Turner would bring the film crews back to the White House for a similar conversation with Reagan, going back and forth until there was understanding and the issues were resolved. That would be the end of the movie – peace between the Superpowers.

It probably would have worked. The movie elite scoffed at the idea, and, undeterred, Turner organized the Goodwill Games with 3500 athletes from 79 countries, and made his own film – a glowing "Portrait of the Soviet Union."

Feeling fairly articulate, I began to write a book about Ted Turner who has always intrigued me. But my words fell flat next to his. After two years of research, I had included wonderful eye witness accounts of those involved with his various crusades. Their words also fell flat next to his. I finally just eliminated the entire book and let Ted speak for himself. It worked.

Here is the amazing story of a child who wrote his life script early, then held to his course until he discovered a New World like Columbus; satellite communications, and with global vision tied the planet together, bypassing borders and time zones.

RIDING A WHITE HORSE

He loved drama, conquering crusades, and adventure on the high seas. At the instruction of his father he read the Great Books and memorized poetry. He wanted to be an explorer or a hero, so was always interested in what you could accomplish if you really tried. Money and power were never his goals. The adventure and the challenge were his beacons.

He spent his life fighting with military skill – lightning strokes, sneak attacks, pre-emptive first strikes – to gain a foothold in business, then to "cure all the evils in the world." He saw bigness as the enemy. The little guy could fight the big guy, but only if he hopped like a rabbit and outran the fox. His business management style was that of leading a charge, brandishing a confederate saber over his head to motivate his troops (staff) against the enemy, known or unknown. His adrenaline rush was contagious.

Working for Ted Turner was a wild roller coaster ride with a mad genius at the helm. The staff was always grateful for his long absences when he was away ocean racing. One staffer said if Ted would just sail around the world for a year he would return to find himself a rich man.

While becoming a force in television, Ted Turner punctured the Iron Curtain with the Goodwill Games at the height of the Cold War. Through the relationships established by the Goodwill Games co-hosted with the Kremlin, he was allowed to set up a CNN bureau in Moscow, breaking the Soviet government monopoly on broadcasting across the U.S.S.R. The Cable News Network became an actual player in world events. The watchful eyes of CNN alerted the Soviet people and world leaders to the attempted KGB coup d'etat on Mikhail Gorbachev, preventing a second Russian Revolution or return of the old-guard dictatorship to the Soviet Union, and also probably saving Gorbachev's life. Through his own travels to hot spots around the world promoting global CNN, Ted Turner helped bring world peace through communications.

"I wanted to tie the world together," he said, "and you know something? It's working."

Just to keep the story alive, we add trophy elk, wild roaming buffalo, Blackfoot Indians and Jane Fonda. We haven't even mentioned the two archaeological digs, the three baby dinosaurs, and the legend of Wowuka. Ted Turner is still riding a white horse – at a gallop.

And here is the rest of the story:

in his own swashbuckling style – in his OWN words.

"I'd like to be Charlemagne. I'd like to ride in on a white horse and cure all the evils in the world."

INTRODUCTION

CRUSADING AGAINST THE BIG THREE NETWORKS

"The networks are run by a greedy bunch of jerks that have hoodwinked the American public, and now I'm riding on a white horse."

BEFORE CNN WAS LAUNCHED IN 1980, TO NEW STAFFERS

"I just want to welcome you all to CNN and wish you the best of luck. See, we're going to take the news and put it on the satellite, and then we're going to beam it down to Russia, and we're going to bring world peace, and all get rich in the process. Thank you very much! Good luck!"

IN MOSCOW 1981, TURNER STICKS HIS FINGER LIKE A PISTOL UP TO THE SKULL OF A COMMUNIST

"Now stick your finger up to my head. There. Now can we talk to each other like this? Of course not. We've got to stop pointing these at each other. You love your children, right? So do we."

TURNER LOSES $26 MILLION ON GOODWILL GAMES

"I'm doing it as a crusade, primarily. I think that strong actions need to take place. I'm not doing it for myself. I'm doing it for the children and their grandchildren and for the people who are getting ripped off – the people who want to live in peace and harmony and want to see our problems solved before we destroy the planet."

ON ONE BILLION DOLLAR DONATION TO THE UNITED NATIONS IN 1997

"What I'm trying to do is set a standard of gallantry. The world is awash in money, with peace descending all over the Earth. We can make a difference in the future direction of the planet."

SPEECH LAUNCHING CNN, 1980

"At CNN we are flying the flag of the United Nations because we hope, with our greater depth, and our international coverage, to make possible a better understanding of how people from different nations can live and work, and so to bring together in brotherhood and kindness and peace the people of this nation and world."

CHAPTER ONE
PUNCTURING THE IRON CURTAIN WITH THE GOODWILL GAMES

THE INAUGURAL GOODWILL GAMES

TED TURNER: (After watching the 1984 Los Angeles Olympics) "Goddurn, Wussler. Why can't we do something like that – but better? – With the Russians?

"The Russians aren't there. And we should have been at their games in 1980.

"Goddamn it, Wussler, that's wrong! Where are the next Olympics? Seoul, Korea? Let's buy the rights. How much do the rights cost? Let's buy the rights and make sure everybody comes.

"Or better yet, I want you to go to Moscow. That's it – I want you to go to Moscow. I want to buy the rights to the whole world for the Olympics with the Russians."

ROBERT WUSSLER (later): "I knew it wasn't possible, but I went ahead. I did what he wanted to do. Sometimes the best way to deal with mad geniuses is to hope they forget their ideas half an hour later, but Turner doesn't always forget.

"I was watching the volleyball final that night and I realized it was a shame we weren't playing the Russians. I don't always agree with Ted, but the next morning I went in and said, 'You're right.'"

TURNER: "Of course I'm right."

WUSSLER RETURNS FROM "IMPOSSIBLE MISSION"

WUSSLER: "One official said, 'The idea of doing something with your country that would honor the last time we were all together in Montreal in 1976, now that is something we would like to do.' They didn't like the Olympic deal, but they'd like to do something in 1986, commemorating the tenth anniversary."

TURNER: "What do you mean?"

WUSSLER: "Well, you know, maybe a multi-sport event."

TURNER: "THEY'D like to do THAT with US?"

WUSSLER: "Yeah."

TURNER: "BOY!" (Doing somersaults, head over heels in coat and tie, through CNN office hallways) "BOY!"

3500 ATHLETES FROM 79 COUNTRIES TO COMPETE IN GOODWILL GAMES

Wussler: "In ten short months, we orchestrated one of the most complicated athletic events and broadcasts in television history. Television engineers, programming people and producers from different countries and all walks of life worked together to build a state-of-the-art broadcast center in Moscow.

"American and international sports federations played a critical role in attracting more than 3000 athletes from a world of scattered athletic interests.

"Sixty-seven independent television stations across America, and foreign television networks around the world, brought the Games into the homes of people on five continents.

"There was one primary source of motivation that helped us all meet our challenge – the lofty purpose of the Games. Perhaps it was best summarized by Mr. Yushkiavitshus during the signing of our agreement: 'It is far better for us to meet on the playing fields than it is for us to meet on the battlefields."

COVERAGE BY WTBS (SUPERSTATION)

WUSSLER (After 20 flights to Moscow): "You know we're going to lose a ton of money on this. Do you want me to cut back? Do you want to cut a couple of days, or do less on the TV production? We're still in a position to cut something. I mean, there's not a lot of fluff. We're doing this with scotch tape and bailing wire but we could use less scotch tape and bailing wire."

TURNER: "No, I don't want you to cut. I want this to look first class."

AT 1986 GOODWILL GAMES IN MOSCOW

"I'm so happy with the way things are going, I'm having a hard time keeping from jumping out of my skin."

OPENING CEREMONIES BENEATH A BANNER READING "FROM FRIENDSHIP IN SPORT TO PEACE ON EARTH"

(At podium) "We can best achieve peace by letting the peoples of the world get to know each other better. Not only will the participants compete together in the spirit of good sportsmanship, but the audiences world wide will see the harmony that can be fostered among nations."

NEITHER NAPOLEON NOR HITLER MADE IT TO THE HEART OF MOSCOW BUT TURNER MARCHED RIGHT UP THE STEPS OF THE KREMLIN

"It's the biggest joint effort between the Soviet Union and the U.S. since World War II."

TALKING TO ATHLETES AND REPORTERS

"How much time we got, huh? What we need is a Big Daddy to take us out behind the woodshed and take a big board and hold us by the ankles and give it to us good."

AFTER DRIVING AROUND HARD TENEMENT DISTRICTS OF MOSCOW

"It's easy to depersonalize people if you've never seen them. When I got into the news business, this became my obligation, to find out about the enemy. I come over here and I say, 'Where's the enemy?'

"I mean, I don't know about all those generals in Red Square on May Day, but then you go into the park and you see all the kids, and you think, 'Geez.' I know things aren't perfect, but I don't see any tin shacks like I did in Sao Paolo or Rio or some of our other friends. This ain't Mexico. What are we fixing to fight these people for? What gives us the right?

"From my office window in Atlanta, I can see hills that still have shell marks from the Yankees' cannons. On my way over here, I stopped at Waterloo and Verdun, to see the battlefields. I got tears in my eyes at the pictures of those bodies lying on the ground. The last madman got 30 million people; 6 million Jews, 24 million others. When's it going to stop?

"Now suppose you got some Captain Kelly, some Russian captain, they're playing pinochle or playing chess, taking dope, or drinking vodka, and they've got a few seconds to figure out if the alarm is real or not, and whether they should deactivate the missile?''

ON U.S. POWER

"You tell me who's got the power. Doesn't seem to be Reagan. When you find out who's got the power, let me know."

LOOKING AT HUGE STATUE IN A PARK

"That's what's wrong with America today. We don't have any heros. Can you name any current American heros?"

PERSONAL MEETING WITH MIKHAIL GORBACHEV INSIDE THE KREMLIN

"I see myself as a citizen of the earth. I don't want to see any nuclear weapons going off over your country or my country. Those are short term victories. Those victories only last twenty-four or forty-eight hours. If we bomb you, it's going to hurt us. If you bomb us it's going to hurt you. Look, I have kids. You have kids. What's their future going to be like?"

CLOSING CEREMONIES

"These games have proven that all the people in the world can cooperate in sports and other mutual endeavors – in a worthwhile and beneficial manner, irrespective of our differences."

EYEWITNESS

Robert Wussler: "I don't think I ever saw Ted happier for a concentrated three-week period of time than I did at the Goodwill Games."

"He poured his heart and soul into the Goodwill Games. He didn't understand that there were people out there who knocked the games as frivolous, unnecessary... and he was truly hurt by the fact that people misunderstood his motives."

TURNER LOOKS BACK ON THE GOODWILL GAMES

"The 1986 Goodwill Games with the Soviets found us with a challenge to demonstrate to the world that we could work toward a mutual goal. Through our shared language and love of sport, we found a means to communicate our goal for peace and understanding. Through our relationship and throughout our telecast we endeavored to show the world that what our nations have in common far surpasses our differences.

"For those of you who were with us during our coverage of the games, we tried to take you beyond the competitors' lives on the field to acquaint you with their customs, and daily routines at home, with their friends and families. We wanted to offer you the opportunity to get to know – and better understand – other people as people.

"The journey to mutual understanding begins with but a single step. It is the hope of Turner Broadcasting System and our Soviet colleagues that the Inaugural Goodwill Games have provided our world a purposeful step on that journey."

IT WAS A GREAT PARTY FOR $26 MILLION

"I didn't do it to make money. I did it to get the two countries on the playing fields again. I could just tell the Soviets were looking to be our friends."

EYEWITNESS

Henrikas Yushkiavitshus, Gostelradio: "To some he looks too idealistic, like Don Quixote, but it is the Don Quixotes who are changing the world, not people who live on stereotyped thinking."

EYEWITNESS

Muscovite hotel waiter: "The Americans will be missed. (Why?) Because Americans laugh all the time. They are always happy and they make everyone around them happy!"

ON ACHIEVEMENT OF GOODWILL GAMES

"I truly believe the Goodwill Games served as a linchpin in bringing the two feuding nations together."

"It was a major mark in history."

"PORTRAIT OF THE SOVIET UNION" ABOUT TO BEGIN FILMING

TURNER: "Don't forget that we're doing a story about a nation of winners!"

IRA MISKIN, Executive Producer: "What are you talking about? This place – on a good day – it's Detroit on a bad day! We're talking about a nation of brilliant history, great literature, and a really oppressive, shitty system. What are you talking about?"

TURNER: "We know all about that but we're talking about WINNERS here. (Running to catch his plane) "Don't forget – a nation of WINNERS!"

"PORTRAIT" IS RELEASED TWO YEARS LATER TO WIDE CRITICISM

William F. Buckley, Jr.: "Turner Broadcasting Company is the outfit that has given us the most innovative and valuable TV programming idea of the decade: round-the-clock news on Cable News Network. That news is reported evenhandedly. But the strangest transformation since the discovery of the transexual operation is what has happened to Ted Turner, the founder of CNN.

"He has become a Soviet apologist. In fact – his stuff is so Red it would embarrass the Daily Worker to publish it...

"When the 'Portrait' was broadcast on WTBS, the first sentence by Washington Post reviewer Tom Shales... was: 'Does Ted Turner have a few thousand acres in the Urals he's trying to unload?'

"Ted Turner was accosted about this travesty on his own 'Crossfire' feature. Pat Buchanan put it to him that this was the most concentrated pack of lies since a broadcast by Goebbels on the Jews. His answer? 'I wanted to go over there and paint a beautiful portrait of the Soviet Union.' Pressed to explain the distortions, he just said it again.

"'Well, that's true,' said Ted, 'That's absolutely true. We went over there to paint a portrait. We painted a portrait, and I'm not going to apologize for it.'"

PERSONAL DIPLOMACY

"Wussler!... I asked Georgi Arbatov (director of the U.S.S.R.'s Institute of USA and Canada Studies) at breakfast yesterday, if we eliminated all our nuclear weapons, would they? And he said YES! I asked him, if we eliminated all of our conventional forces in Europe, would they? And he said YES!"

LOOKING OUT OFFICE WINDOW

"I love this view. I'm near all my friends. I've got former President Carter a few miles away over there.

(Then pointing northward) "And I've got Gorbachev just a few miles farther away, over there.

(Peering through a kaleidoscope): "When I was younger, racing all over the world and having a ball, I didn't think about the world situation. I used to think everything was fine. So then, here I was in 1980, starting CNN – this global, twenty-four-hour-a-day news network – and I decided I needed to find out what was happening in the world. You know, what was REALLY happening.

"I started studying the world, and I got together with a group of experts and politicians who understand the planet better than I do. It's a crazy idea, I know, but I decided we had to take better care of our planet, because in taking care of our planet, we might be able to save ourselves. At the rate we're going, man is the most endangered species on earth. If we don't become extinct by exhausting the planet's natural resources, then we'll destroy ourselves by nuclear war. And it doesn't have to be that way.

"The earth has about five billion years of life before the sun depletes itself. We can be around that long if we don't destroy ourselves through waste or war. Why rush it?

"In 1985, I started this group called the 'Better World Society.' We're kind of like a mini United Nations. Lester R. Brown, a Better World Society board member, is the head of the Worldwatch Institute, in Washington, and his group puts out a book every year called "State of the World," which is an inventory of life on earth. You know, how the world's inhabitants are doing – man and the trees and the animals. And we use this information for good, common sense global policy.

"That thinking extends to nuclear weapons, too."

(For Christmas that year, Ted sent a gift-wrapped copy of "State of the World" to every member of Congress and every Fortune 500 executive.)

"We need to totally disarm, because, just as with guns, if you've got nuclear weapons you're eventually going to use them."

GORBACHEV INVITES TURNER TO SPEAK TO STUDENTS AT HIS OWN ALMA MATRE

"Last year I spoke to Moscow University and I gave them something to really work for. You know we've got the year 2000 coming and we've got a challenge ahead of us that mankind's never had before. We're really on the verge of having peace on this Earth. We really are. There is not a major war going on in the world. Wars don't work. We couldn't win in Vietnam. The Soviets couldn't win in Afghanistan. Iran and Iraq were just losers. Everyone's losing in Lebanon. Everyone's losing in Northern Ireland. It's really time that we put that behind us. And that's really something...

"If we had peace – that's something we've never had – Peace. And we really need to have peace, not only man with man, but man at peace with the environment. A tree doesn't have much of a chance, the forest doesn't have much of a chance against bulldozers and chain saws. So we really need peace with the environment, too, and that's a possibility also.

"And I'd like to see a movement going. The most exciting thing that we could do – the last time the calendar was turned over was 2000 years ago, when we had A.D. and B.C. Christianity's had two thousand years to solve the world's problems and they're just as big now as they've ever been – and the missionaries have either been killed or eaten in the Third World. So Christianity is not going to take over the world – so why don't we start over?

"Why don't we aim during the next ten years to have peace on Earth? And in the year 2000, turn the time back to zero. And let it be B.P.

and A.P., 'Before Peace' and 'After Peace.' That could be the greatest honor we could bestow on our generation. So if we do that, then people will BE here 2000 years from now.

"If we continue on the course we've been on...

"The most important thing we can do is see that our children live a decent life – and live to be seventy years old like we hope to do."

CRUSADING AT A DINNER PARTY

"Might does not make right. Right makes might.

"I'd rather be the one to get blown up than the one that blows up. I have to love everybody. Like Jesus.

"I've made my peace with the Soviets. They're not MY enemies. We're all brothers. No one is going to drop a bomb on his brother, except some dumb, mean son of a bitch and those are the ones I don't want anything to do with. I expected Castro to be some horrible person, but he was a good guy. I walked the streets of Moscow and saw mothers with their little children. There weren't KGB agents on every corner. I found their leaders very reliable.

"I really wonder if there ARE horrible people. Anyone who kills or hates or hurts other people – it's all the environment. Basically people are damn good, that goes for Catholics and Jews and that goes for Communists. Even prisoners are nice guys. A convicted murderer presented me with a plaque for speaking in prison.

"I have a 2000 pound bison that killed a horse, and I feed him out of my hand. If you can get along with animals by being nice to them, you can get along with human beings.

"Capitalism didn't intend for the workers to starve. If the rich were smart enough to eliminate poverty, and make everyone happy – then they'd have more mouths to sell to. But when a few wealthy persons are pigging out in a world surrounded by poverty, that's a very dangerous situation...

"I drive a Toyota, and have since 1974. We're the most wasteful materialists on earth. We bully smaller countries. We're living higher on the hog than we ever have – we have three cars; not one house any more, but two houses. Five percent of the world's population is consuming 35 percent of the world's resources... And the average American eats two and a half times more meat, and...

(When asked about all his land holdings) "I guess I am acquisitive, but that's my capitalist background. Those who make greater contributions deserve greater rewards. Generally, people believe that leaders are entitled to live at a better level."

ON GOODWILL GAMES

"I think it was a success in '86. It was not a financial success, but we got the U.S. and the Soviet teams on the field for the first time in ten years. And I think as a result of that, the Soviets will be on the field next year in (the Olympics) in Korea, and they might not have been – because South Korea is our strong ally, and there are no diplomatic relations between South Korea and the Soviet Union.

"By our bringing the U.S. athletes to Moscow, where the boycotts started in the first place, I believe we changed the Soviet view of athletics at the Olympic level, with the United States, and we did a lot to start a thaw in U.S./Soviet relations.

"And I think it was a tremendous success. Maybe we'll only break even or lose a little money in 1990, but by 1994 we should make a lot of money.

"It's just got to be a good idea in the long haul. We don't feel like we're competitive at all with the Olympics, but we feel like we're supplemental to the Olympics... But there is no reason why the Goodwill Games can't be reasonably as successful in time as the

PUNCTURING THE IRON CURTAIN

Olympics are. We think we have a very, very valuable franchise that we are building up.

"All the leaders in the cable industry came as our guests to the Goodwill Games in Moscow. Our new partners (in TBS) are those same guys that we spent a week over there with touring Moscow and Leningrad. They had a ball."

"It's a new challenge – to get along with partners. The first time I ever had partners was for the Goodwill Games. We were partners with the Soviets. And even though there was some hard bargaining and negotiations, it worked out just beautifully. Now we really are partners with them, on-going, and it's really exciting."

"I believe that in order to solve the problems of today – we have to work together. We've got to strengthen the United Nations. We're going to have to learn to get along with everybody. And that's a big difference, that is a major change. And in order for me to be able to espouse that philosophy and promulgate it, I have to have partners of my own and learn how to get along with them.

"When you've got partners, you have to spend some time listening. You can't be talking all the time."

1990 SEATTLE GOODWILL GAMES

LOSES $44 MILLION

"The world is a lot less dangerous...

"Who cares why? If the Goodwill Games had something to do with it, so much the better. But you've had Pepsi-Cola there for a long time, and medical doctors and Ted Hesburgh and John Denver.

"The Soviet leadership feels I was friendly with them all along, but it's hard to ask me about what we've accomplished. I think big. I'm a big thinker, but there's no way I could have predicted what has happened... (the toppling of the Soviet Union and the Berlin Wall).

"These games will lose more than half of the $26 million they lost in 1986, but I look at it as investment spending.

"But the battle is still not won. Relations between the countries are better, but we still have 30,000 missiles on hair trigger. They haven't got rid of a single missile in most places. They still haven't done much. It's like a relationship with your wife. You can't take it for granted."

THE 1994 ST. PETERSBURG GOODWILL GAMES

EYEWITNESS

"Ted Turner moved briskly about the ornate corridors of the Hotel Grand Europe as easily as if he were Tedski Turnerski, exchanging greetings and handshakes with passersby who obviously enjoyed his attention."

HOW HAD TURNER RISEN FROM UHF OBSCURITY AT SUCH A DIZZYING PACE?

"Satellite. That was the big step.

"I'm like the bear that went over the mountain to see what he could see. One thing opened up another, and I kept moving on. Remember,

I went from a small sailboat on Lake Lanier to winning the America's Cup."

"We've had some complaints from the board about the Games losing money, but as the losses have gone down, so have the complaints. Our directors are for the Goodwill Games all the way.

"We are fully committed to New York in 1998. Governor Cuomo will be here for the closing ceremony and the passing of the torch."

1998 NEW YORK GOODWILL GAMES

$5 MILLION PURSE OFFERED TO ATHLETES: GAMES ARE TO BE A BENEFIT FOR UNICEF AND THE BOYS AND GIRLS CLUBS OF AMERICA

Mario Cuomo (then Governor of New York): "We welcome this opportunity to present the Goodwill Games as the important vehicle of understanding that they are, not only in 1998 but also on into the next century as well. The Goodwill Games represent a standard that we strive for in New York, and together we believe we can convey this sense of community to the world."

(See appendix for events, dates and locations.)

CHAPTER TWO
A BORN CRUSADER

AGE 10, CHANTING MEMORIZED POETRY AT SAVANNAH BEACH, SC

> "Ay, tear her tattered ensign down!
> Long has it waved on high
> And many an eye has danced to see
> That banner in the sky.
>
> O better that her shattered hulk
> Should sink beneath the wave!
> Her thunders shook the mighty deep,
> And there should be her grave:
>
> Nail to the mast her holy flag,
> Set every threadbare sail,
> And give her to the god of storms,
> The lightning and the gale!"
>
> "Old Ironsides"
> Justice Oliver Wendell Holmes

(This was the poem that saved the USS Constitution from being scrapped. In 1997 CNN gave extensive coverage to the Bicentennial restoration and sailing of this famous frigate, built not from steel, but hand hewn Southern live oaks.)

SELF-EDUCATION

"I was a vociferous reader when I was young."

READING AT HOME

"In the summers, because my grades were not good, my father made me read the Great Books. I never considered not doing it because I was instructed with wire coat hangers when I didn't get them read."

"My father had an idea of what I should do in life and the way I should live, right down to the finest detail."

"My father put the screws to me early. If he hadn't, I never would have survived. My father made me a man."

READING AT SCHOOL

(At McCallie School) "I started with the fables, Aesop's first, then the others. (He went on to the stories of the Gods in both Greek and Roman mythology.)

"I was interested in one thing, and that was in what you could accomplish if you really tried. So I looked around for guys who had tried. We had four hours of study hall every day and I'd blast through the uninteresting part of my homework, particularly in the early years, and be content to get C's. But there were always monitors coming up and down the halls, and since you had to be busy, I could read anything.

"I read about the sea because I had a little sailboat of my own. I read C. S. Forester's books, and Nordhoff and Hall about ten times – 'Men Against the Sea,' 'Mutiny on the Bounty,' and 'Pitcairn's Island.' I read about the War of 1812 and about the 'Constitution,' you know – the ship. I remember reading the story of the Marines at Tripoli.

And then I would go on to the dreadnoughts in World War I, and then – when I'd gone through that, on to World War II.

"My interest was always in why people did the things they did, and what caused some people to rise to glorious heights, like the Macedonians did under Alexander the Great. Alexander decided to go farther than anyone had tried to go: there were no limits to his imagination.

"Well, that stuff just knocked me cold. I used to cry over those stories. Alexander the Great was far from perfect, and yet he accomplished all that he did because he wanted to and worked at it all the time. And his men loved him, even though they mutinied after fifteen years on the hike. It was so exciting going to new lands all the time. They had trophies, and I always loved trophies – I still do. You got gold and silver when you won a battle – something tangible."

CHILDHOOD HEROS

"I wanted to be a kind of knight in shining armor, and at that time, Alexander the Great, Napoleon, and George Washington – the military leaders – were the heros I looked up to the most."

DISCOVERING NEW WORLDS

(Just before launching CNN): "When I was a kid I was really upset there weren't any new worlds left to discover. But this new world is better than anything Captain Cook ever dreamed of. I'm so proud of all this I just can't stand it. I've got at least five lives to lead.

"I've got so many lives to lead there just isn't enough time to do it all. That's why I work so hard and spend so much time on my business. I've got so much to do I can't believe it. But damn, it's all so much fun..."

AGE 9, GEORGIA MILITARY ACADEMY WAS NOT SO MUCH FUN

"I got beaten up all the time, or at least I had to fight all the time. I don't know what it was. Yes, I do know what it was: the other kids thought I was a show-off and a smart-ass."

"It was pretty rough. I was from Ohio. I was a Northerner, and I didn't enroll until about six weeks after the regular term had begun, so I was coming in late...."

"Someone had started a rumor that the new boy – the one with the Yankee accent – was bad-mouthing General Robert E. Lee, and soon I was running for my life, pursued by a bloodthirsty platoon of about forty cadets brandishing a rope and screaming 'Kill the Yankee bastard!' I hid in my locker for about four hours while they were out there. They were in a rage. If they got me, they would have killed me. I was in terror."

AGE 11, MCCALLIE MILITARY SCHOOL

"I was certainly the only boarding student." (Of that age)

"At first I was just a terrible cadet. I did everything I could to rebel against the system, although I believed in it from the beginning. I was always having animals in my room, and stuff like that, and getting into trouble one way or another, and then having to take the punishment like a man.

"But at least everybody was equal there, which is what the military system does. I went through a lot, but it changed me, and after a while I shaped up."

EYEWITNESS

Houston Patterson, teacher: "He wasn't disliked. He just went on his own little beacon."

BIBLE CLASS

TURNER: "I don't think I'd like to go to Heaven."

BIBLE TEACHER: "Why not?"

TURNER: "Well, I just can't see myself sitting on a cloud, playing a harp day in and day out."

BIBLE TEACHER: "Well, tell me, Ted, what is the most perfect way to spend the day that you can imagine?"

TURNER: "Being with my best friend on a golf course with some money in my pocket and no one before us, no one behind us, just ourselves on the greens. Then we play another round. Then we go swimming. Then we go to the clubhouse and get a sandwich."

BIBLE TEACHER: "Ted, that's wonderful. It says here in '1 Corinthians,' 'Eye has not seen, nor ear heard, nor has it entered the heart of man, the things (plural, he stressed) that God has prepared for those that love him – to me – Ted, that means that either there will be the best golf course that you have ever seen, or something so much better you'll forget all about golf."

TURNER: "Well that's a pretty clear-cut choice, if that is true, so I was thinking about all these people in other parts of the world where they don't have access to Christianity, they – you know, I'm just a little kid – you know. I thought, Geez, these people are all going to Hell, and I thought I'd go out and try to help as many of them as I can."

SERIOUSLY PLANNED TO BECOME A MISSIONARY

"In those days I became very religious."

"I wanted to go to some little Timbuktu and try and get the people to help me build a church, and – you know – convert them... to save their souls, help them save their souls."

RESOLVES TO BECOME THE BEST

"I had worked hard at being the worst cadet and decided I wanted to be the best, and I saw that it could be done if you worked at it. By the time I got out I had accomplished something. I was the Tennessee State debating champion. I beat a girl in the finals. She broke down into tears because I challenged the basic premise being debated. I was named best dressed cadet, and in my Junior year I got to be an officer and an inspector – and then – it was me who went around looking for dirt and giving other people demerits. But I always tried to be fair when I finally got into a leadership situation."

"For several years I was absolutely the worst cadet in the place. I didn't do anything, and what I did do got demerits. Then I turned it around. I'd been the worst cadet, and I determined to be the best. I became a believer. I ended up with 'best cadet' honors. When I left there, I cried. It was such a perfect place."

"I've always felt that McCallie was kind of like my second home."

REVISITING MCCALLIE

"I loved this school a lot. It did a lot of good for me – a lot of times you don't appreciate things as much when you are there as when you have the opportunity to look back on them from a number of years.

I thought the education here was terrific. I learned to think here for the first time...

"I didn't really set out to make a lot of money, I really didn't, and that has not been my major motivation. I just wanted to see if these things could be done... I feel that had I been born four or five hundred years ago, I would have wanted to be an explorer, like Columbus or Magellan. In fact, I probably would have preferred to do that... but I did get to be a kind of pioneer or explorer in the television business... With technological change coming as fast as it is in virtually every industry, it just comes so fast it is hard to keep up with it. It's like a run-and-gun basketball game. You've got to keep your eyes open, because the ball might be coming at you from any direction and you want to get those uncontested lay-ups.

"That's basically what I did. I just broke away from the field in television and was out on the corner and they got the ball to me and I was able to take three steps and dunk the ball... It wasn't hard at all."

AFTER A SEVEN YEAR BATTLE WITH SYSTEMIC LUPUS ERYTHEMATOSUS TED'S SISTER DIES AT 19

"She was sweet as a little button. She worshiped the ground I walked on. A horrible illness."

FAMILY HIRED 'ROUND THE CLOCK NURSES

"She came out of that coma with her brain totally destroyed. It was a horror show of major proportions. A padded room. Screaming day and night. It was something right out of 'Dark Shadows.'"

POEM WRITTEN WHILE MARY JEAN WAS SUFFERING

"For Lack of Water"

..."Awful sin had he committed?
    Wrath of God had he incurred?
As he plodded on he wondered
    But his mind and thoughts were blurred.

One could see his strength was failing
Two more steps and he'd be through."

BROWN UNIVERSITY

"I'm Ted Turner from Savannah. I'm the world's best sailor, and the world's best lover."

"There was a lot of bull at Brown. It mattered who your father was, how much money you had, what your clothes looked like. I was used to a certain directness, and there was very little of that."

"Brown was too much like prep school. I was expecting it to be more mentally enlightening. My professors just didn't motivate me."

EYEWITNESS

Peter Dames: "Basically, we both liked to get drunk and chase women, but we had a couple of things in common that were important. We had both come to Brown fresh out of five years of military school, me at Manilus in New York State and Turner in Chattanooga, and we had had parallel careers. We had both been the worst possible cadets at first, figured out the system, and gone on to run the places. So the metamorphoses were the same.

"This also meant we were both totally unprepared for Brown's social life. Everybody else had gone to Choate or Lawrenceville, or some other fancy prep school, and they were all presidents of their class or captains of their football teams. I mean nobody was an average guy.

"So Turner and I had to look around for something to excel at, and we settled on drinking and lechering. We really worked hard at it, too, and we excelled. We became legends in our own time. If he got some money, or I did, we would immediately invest it in improving our reputations. You have to understand that for the other guys, this was nothing new. They had done it all at Choate and Lawrenceville. But for us, suddenly there were no lights out at ten o'clock, no bed checks, no inspection. You could drink and screw all you wanted. And we wanted. If Turner got thrown in jail, I would bail him out. One time I tried to bail him out and they locked me up too."

ON BROWN

"I was really happier at military school. I didn't like it when there were no rules... I never did graduate, but I learned a lot and I accomplished some things. I was the captain of the sailing team, and also the coach – we had student coaches then. I spent all my time at the Brown yacht club, working on the boats and traveling to college regattas instead of studying. But you never know. If I'd gone to Annapolis I would have done great, and I'd be an Admiral by now.

"But no one would have heard of me."

FATHER'S INFLUENCE

"My dear son,

"I am appalled, even horrified, that you have adopted Classics as a Major. As a matter of fact, I almost puked on the way home today. I suppose that I am old fashioned enough to believe that the purpose of an education is to enable one to develop a community of interest with his fellow men, to learn to know them, and to learn how to get along with them. In order to do this, of course, he must learn what motivates them, and how to impel them to be pleased with his objectives and desires.

"I am a practical man, and for the life of me I cannot possibly understand why you should wish to speak Greek. With whom will you communicate in Greek? I have read, in recent years the deliberations of Plato and Aristotle, and was interested to learn that the old bastards had minds which worked very similarly to the way our minds work today. I was amazed that they had so much time for deliberating and thinking, and was interested in the kind of civilization that would permit such useless deliberation. Then I got to thinking that it was not so amazing after all they thought like we did, because my Hereford cows today are very similar to those ten or twenty generations ago.

"I am amazed that you would adopt Plato and Aristotle as a vocation for several months when it might make pleasant and enjoyable reading to you in your leisure time at a later date. For the life of me, I cannot understand why you should be vitally interested in informing yourself about the influence of the Classics on English literature. It is not necessary for you to know how to make a gun in order to know how to use it. It would seem to me to be enough to learn English literature without going into what influence this or that ancient mythology might have upon it.

"As for Greek literature, the history of Roman and Greek Churches, and the art of those eras, it would seem to me that you would be much better off learning something about contemporary literature and writings, and things that might have some meaning to you with the people with whom you are to associate.

"These subjects might give you a community of interest with an isolated few impractical dreamers, and a select group of college professors. God forbid!

"It would seem to me that what you wish to do is establish a community of interest with as many people as you possibly can. With people who are moving, who are doing things, and who have an interesting, not a decadent, outlook.

"I suppose everybody has to be a snob of some sort, and I suppose you will feel you are distinguishing yourself from the herd by becoming a Classical snob. I can see you drifting into a bar, belting down a few, turning around to the guy on the stool next to you – a contemporary billboard baron from Podunk, Iowa – and saying, 'Well, what do you think about old Leonidas?' Your friend, the billboard baron, will turn to you and say, 'Leonidas who?' You will turn to him and say, 'Why Leonidas, the prominent Greek of the Twelfth Century.' He will in turn say to you, 'Well who the hell was he?' You will say, 'Oh, don't you know about Leonidas?' and dismiss him, and not discuss anything else with him the rest of the evening. He will feel that you are a stupid snob and a flob: and you will feel that he is a clodhopper from Podunk, Iowa. I suppose this will make you both happy, and as a result of it you will wind up buying his billboard plant.

"There is no question but that this type of useless information will distinguish you, set you apart from the doers of the world. If I leave you enough money, you can retire to an ivory tower and contemplate for the rest of your days the influence the hieroglyphics of prehistoric man had on the writings of William Faulkner. Incidentally, he was a contemporary of mine in Mississippi. We speak the same language – whores, sluts, strong words, and strong deeds.

"It isn't really important what I think. It's important what you wish to do with your life. I just wish I could feel that the influence of those oddball professors and the ivory towers were developing you into the kind of man we can both be proud of. I am quite sure that we both will be pleased and delighted when I introduce you to some friend of mine and say, This is my son. He speaks Greek.

"I had dinner during the Christmas holidays with an efficiency expert, an economic advisor to the nation of India, on the Board of Directors of Regents at Harvard University, who owns some 80,000 acres of timber land down here, among his other assets. His son and family were visiting him. He introduced me to his son, and then apologetically said, 'He is a theoretical mathematician. I don't even know what he is talking about. He lives in a different world.' After a little while I got talking to his son, and the only thing he would talk to me about was his work. I didn't know what he was talking about either, so I left early.

"If you are going to stay on at Brown, and be a professor of Classics, the courses you have adopted will suit you for a lifetime association with Gale Noyes. Perhaps he will even teach you to make jelly. In my opinion, it won't do much to help you learn to get along with people in this world. I think you are rapidly becoming a jackass, and the sooner you get out of that filthy atmosphere, the better it will suit me.

"Oh, I know everybody says a college education is a must. Well, I console myself by saying that everybody said the world was square, except Columbus.

"You go ahead and go with the world, and I'll go it alone... I hope I am right. You are in the hands of the Philistines, and damnit, I sent you there. I am sorry.

"Devotedly, Dad"

TED DUTIFULLY CHANGES MAJOR TO ECONOMICS

"When I got into economics, I began running into commie professors who thought everybody ought to work for the government. I was opposed to that and defended the free enterprise system to the extent I almost flunked the course. To me the capitalist system is still the best way to get things done.....What a great system!"

EXPELLED FROM BROWN IN FOURTH YEAR

"At Brown I was a rebel ahead of my time. I got thrown out of college for having a girl in my room. Today they have girls and guys living in the same dorm."

CHAPTER THREE
LIGHTNING STROKES, SNEAK ATTACKS

JOINING FATHER'S BILLBOARD BUSINESS

"I worked fifteen hours a day, six-and-a-half days a week. We doubled our sales in two years. It was phenomenal."

MOTTO

"Early to bed, and early to rise, work like hell, and advertise."

EYEWITNESS

Judy Nye Turner, first wife: "He thought billboards were great. They were visible: people would see them again and again.

"Unlike a magazine or a newspaper which you throw away, a billboard is in the same place day after day. Then, too, the profit margin was wonderful in billboards. Once a board was constructed, you just had to keep the weeds cut, and paint it every now and then to keep the rim of it nice. And you could get pasters for cheap. It wasn't really a skilled labor job. So as advertising, the mark-up was just wonderful."

EYEWITNESS

Edwards: "Ted traveled with me for about six months but you could tell he was a natural from the git go. Ted was one of the greatest salesmen in the world. Still is – just like his father. Those two, they were so much alike they couldn't stay in the same room ten minutes together without arguing over the best way to do this, or do that. But either one of them could charm a rattlesnake."

EYEWITNESS

Judy Nye Tuner: "I think his father was so proud of Ted then. He thought the world and all of him. He wanted to create a superman. I really believe that was to be his legacy. His life was going to be through Ted."

EDUCATION FROM FATHER

"Driving in to work, he told me about the tax laws, amortization, depreciation, sales, management, construction. He told me how he got started, what happened in competitive situations, how he lost business and how he got it....

"All my life, I have had this gnawing pain that I might not succeed. It is only in the past four or five years that I have put that ghost to rest."

FATHER'S WORDS

"You know, you have the opportunity to do anything that you want to do. You just have to take time out to think about things, and then you have to do them."

"People think I'm a crazy man. But my father, he REALLY was the crazy man. He lived hard, played hard, did outrageous things. I mean, he used to go into bars and get in FIGHTS and stuff."

TED'S FATHER HAS A BAD BOUT WITH ALCOHOL AND DEPRESSION

"Looking back, for about six months he had this terrible state of depression."

"I thought he was having a nervous breakdown. I begged him to stop working – to take some time off – but he wouldn't... Nobody had any control over him. He was his own man. And I didn't even know what the problem was at the time."

"My father could be absolutely charming or he could be a horse's ass. He could be the kindest, warmest, most wonderful person in the whole world, and then go into a bar – get drunk and get in a fist fight with the whole place.

"He was a rugged individualist. It's an old phrase but it describes him well, because he was in fact a throwback to the past. He didn't have a whole lot of fear, but sometimes he did have remorse, remorse for the things he had done. He had a bad habit, and it was that he would always say exactly what he thought, without being diplomatic at all. That got him in a lot of trouble – along with drinking too much. I've tried to learn from that. Maybe I tend to be outspoken myself – but he was so outspoken I saw it cost him a lot of friends, and it cost him a lot of money, too. He would have gone a lot further if he hadn't been so controversial.

"I loved him, I know that. We loved each other, and yet we were so cruel. He was a hard man, and I tried to please him, although I didn't a lot of the time, and we had terrible, terrible fights.

"It was after one of those fights – we disagreed how the business should be run – that he blew his brains out."

RIDING A WHITE HORSE

"Dad just got anxious. He got emotional. He was working too hard, drinking too much, popping pills, sick all the time. The pressure finally got him. Two years before, he said he wouldn't mind if he died. I knew then."

"He put a bullet through his head with the same gun he taught me to shoot with. At the end, the banks wouldn't even honor the check for his funeral."

"It was devastating."

FATHER HAD BOUGHT ANOTHER BILLBOARD BUSINESS WHICH INCLUDED ATLANTA TERRITORY, LOST HIS NERVE, AND SOLD IT BEFORE HE TOOK HIS LIFE

"My father was halfway to the big time. I think he saw me as the only hope for the family to go all the way... It took my father all his life to get to Atlanta. I wasn't going back. No way. It was a hard, bad road."

WITH ACCOUNTANT ON THE NIGHT AFTER THE FUNERAL

TURNER: "I want to run the business my father left me."

IRWIN MAZO: "Come on, Ted. Forget it. Why don't you go sailing? That's your first love anyhow."

TURNER: "No, my father wasn't in his right mind. I have to do this. My father didn't leave me all this money just to go sailing. He left it to me to run the company."

IRWIN MAZO: "But you don't have the line of credit your father had. You're only twenty-four, and the banks don't know you.

Besides, we've got estate tax problems, plus certain stipulations made by General Outdoor Advertising – which is buying. Even the things which were easy for your father wouldn't be available to you."

TURNER: "I still want to do it."

IRWIN MAZO: "I don't think you can."

DETERMINED TO FOLLOW HIS FATHER'S DREAM

"He was scared. I kept telling him we could pull it off - that it wasn't a problem - and he kept saying 'Hell, no. I'm going to get rid of it.' He sold it for a pittance. I mean – he sold it for what he paid, plus a little – but not what it was worth. I'm going to take these people to court to break this contract if they insist – if they won't settle.

"I'll go to court and testify that my father was emotionally unstable, that he was on an alcoholic binge, that his thought processes were unclear – that he wasn't capable of making this kind of decision...

"I am perfectly prepared to get on the stand and say, from my last meeting with him, that my father was very depressed, that his chronic alcoholism was very much in evidence, that he was not in his right mind, and that he was not capable of making a lucid or intelligent decision."

DETERMINED TO REGAIN PORTION OF BUSINESS HIS FATHER HAD SOLD TO NAEGELE

"I was sad, pissed, and determined. I had to get that company back.

"I knew Mr. Naegele was my father's best friend, and I thought I'd just call him up and it would be fixed. But it was still cold in Minneapolis, so Naegele was out in Palm Springs. I jumped on a plane and went out there. 'Just tear up the purchase agreement and let me keep my father's company. You won't be sorry. You don't really want to do this anyhow.' The answer was no."

USING TACTICS OF HIS MILITARY HEROS

"Lightning strokes, sneak attacks! Hit them before they know what's happening.... Don't give them a chance to regroup. That's the only way it will work. That's the only way a little guy can beat a big guy. No holds barred. Don't be afraid because you're little and you're afraid and it looks like you haven't got a chance. The rabbit can get away from the fox, but he'd better get on his hind legs and hop.

"I was actually on my way back to Atlanta when I grabbed a phone in the airport and scheduled a meeting of our lease department. That's the department that has all the contracts for the leases on billboard locations which in that business is the most important asset you really have. The meeting was set for that night, and when I got back, I just hired the whole department, and put them on the Macon payroll.

"See the deal with Naegele was – we were just supposed to operate the Atlanta company temporarily until they took over, but because all they'd signed was a purchase letter there was no actual non-compete agreement.

"So the next morning my new employees and I went out and started transferring the Atlanta leases to the Macon company. Just a little paperwork, we'd say if anybody asked. I delayed as much as I could with Mr. Naegele so that when they finally came down with the closing documents two weeks later I presented them with a fait accompli.

"I told them I had already hired the entire lease department of the company they were supposed to be buying, and I'd already jumped the leases – that's when you go out and get a new lease for another company. It's sabotage. If you ever want to steal a franchise that's how you do it. 'Furthermore,' I said, 'I can delay another two weeks and then I'm going to burn all the records. You're going to have

LIGHTNING STROKES, SNEAK ATTACKS

nothing but a disaster.' Boy, did they get back on the telephone to Minneapolis.''

"I also threatened to build billboards in front of theirs. My father wasn't in his own element when he sold out, but I'm in mine."

"The big cheeses out there had an idea. They said they would give me two hundred thousand dollars if I would give them all the leases back and be a nice boy. Or, they said, if I'd give THEM two hundred thousand, then they'd release me from the contract. They said I had thirty seconds to decide – but I didn't even take the thirty seconds.

"I said, 'That's a fair deal, I'll give you the two hundred thousand.' Well, they all fell over backwards. There were actually three guys involved in the deal and all three of them were in the 90 per cent tax bracket. They only owned the thing for a couple of weeks, so it was a real short term deal, and all the two hundred thousand would have been ordinary income. They could see it all going to Uncle Sam and they didn't know what to do.

"They'd given me thirty seconds but when I said 'yeah,' they ran away to think some more."

TO ACCOUNTANT

"By the way, Irwin, where do we get the two hundred thousand dollars?"

BUYS COMPANY BACK FOR NO MONEY DOWN

"I wound up convincing them that they should take stock in the company, so later on they would get a long term gain. In the end, we didn't even have to pay the two hundred thousand.

"But boy I had plenty of other debts. There was a $600 thousand payment coming up in six months to the First National Bank of

Chicago. I was only a kid, but I learned how to hustle. I went out and convinced the employees of the company to buy stock in it. I sold off all the real estate I possibly could, I sold my father's plantations, I borrowed against our accounts receivable, and I squeezed the juice out of everything."

TRIBUTE TO FATHER IN OUTDOOR ADVERTISING NEWSLETTER

"I want to thank you for your kindness to my father while he was alive. He loved this business and spent his whole life trying to make it better in his own way. He may have been independent in some ways, but he was honest, fair, and never spared himself.

"We have amicably concluded recent negotiations with the Naegele Companies to the satisfaction of all concrned. We will retain Atlanta, Richmond and Roanoke. I want everyone associated with the business to know that Turner Advertising Companies fully intend to continue forging ahead, providing the finest in both plants and services.

"We intend to cooperate on all reasonable joint efforts with the dedicated men and women in this industry in the future growth of outdoor advertising, and we pledge to our advertisers our intent to make dollars that are spent with us reap a bounteous harvest in the form of increased sales."

EYEWITNESS

Judy Nye Turner: "He was going to hold it all together. I mean, this was the big opportunity.

"It might have been the only opportunity he was going to have. His dad had been waiting for this for years. So why would Ted want to let it go? No way. I don't think there was ever a doubt in his mind that he was going to accept (just the money from the sale). I don't think there was ever any question....

"I mean, he was groomed for this. There was no way Ted was going to let that go."

EYEWITNESS

Irwin Mazo: "Right off, Ted created a sense of paranoia within the company, a sense that we were the little guys fighting for our lives against some unknown big guys. That made everything seem a lot more important than it probably was. After all, we were not doing badly. And Turner Advertising was the biggest billboard company in the South. One of the biggest in the country actually. Ted compounded the sense of danger, though, with all kinds of subterfuge. He insisted on making his telephone calls to me on outside pay phones. He wanted people to believe our phones were tapped. Perhaps it would distract them from how well the business was doing."

MOTIVATING THE TROOPS

"Stick with me. If I make it, you're going to make millions. Stick with me. We're going places...

(About naysayers) "Those dumb bastards! We're going to make something out of this thing. We're going to make this thing work. If we make it you're all going to be rich. And if we don't – well, what the Hell! What've you lost? You'll be young enough. You can do something else – go down to Florida and sell advertising in the Yellow Pages. What the Hell!"

"If things get really bad, I can always kill myself."

OUTDOOR ADVERTISING CONVENTION SPEECH

"I'd like to point out that we're tenacious. We don't accept marginal locations. We pick the location that we want, then keep going back until we get it. In fact we literally hound the people to death in a nice way. I've had people throw up their hands after forty hours with them on about thirty calls, and say 'O.K., O.K.'"

ON EXPANSION

"What you do is you get a bank, and you borrow all you can borrow. You borrow so much that they can't foreclose on you....

"I don't have any money. I'm just going to keep borrowing. You know, it's like a Ponzi scheme. The whole thing is a Ponzi scheme."

EYEWITNESS

Judy Nye Turner: "The dream was just to build on the dream until you can't go any further."

EYEWITNESS

McGinnis: (A Greenville company began putting signs in Ted's territory because he was away sailing so much) "Oh, boy, was Ted mad! But the way he saw it, if the other guy was trying to cut into the action in Charleston, he must not be busy enough in his home town. Because – if he was busy, see – he wouldn't be fooling around in Charleston.

"So, we sent some guys to keep him busy. Actually, we sent an armada of Turner people and trucks to Greenville, and we built a whole bunch of billboards overnight. It was a lightning attack, and it brought the guy to his knees. Bang, he went down. We ended up selling him – at a premium price – all the boards we built in Greenville, and got from him everything he'd put up in Charleston."

BILLBOARD BUSINESS STABILIZES

"Hell, after about four years in the outdoor business, during which time I was racing my boats most of the time, I could've retired. We had a lot of fun, getting up at five in the morning to go out and put up a new sign before traffic got too bad. We were like Michaelangelo painting the Sistine ceiling – except you could stand up – you didn't have to lie on your back.

"One night we were painting this 48-foot board of the Coppertone girl, and the guys forgot to put on her bikini. It was a work of art – the chest and the crotch were perfect. I always regretted making them put the bikini back on before she hit the street.

"But after a while I felt the billboard industry had matured, and you always have to move ahead."

EYEWITNESS

Jim Roddy, staff: "Despite all his other remarkable achievements, his salvaging of that company, with real big sharks biting at his heels, was the first proof of the pudding. He rallied his people and he worked like hell, and it may have been his finest hour."

CHAPTER FOUR
RADIO AND TV – TALKING BILLBOARDS

LIVING IN ATLANTA,
TED MARRIES JANE SMITH

Janie: "My own babies came along one after another and I was left alone. I was miserable a lot of the time. Every chance I got I would load up the car with cribs and diapers and stuff and drive home to Birmingham.

"Three times Ted was away from home sailing over Christmas. One Christmas I was so pregnant I couldn't even go home and my parents couldn't come to Atlanta for some reason, and so I just stayed in Atlanta. I cried and cried.

"And I was always on call. After the first transatlantic race, he telephoned from Denmark and I flew over to meet him. I had to wait for the phone to ring, because we never knew how long a race might take, or where they might end up. Rhett was just a tiny baby, so I left him with my mother and took off. We didn't even come right home. Ted had arranged to be in another regatta in Copenhagen, and we stayed for that. King Constantine of Greece was sailing – I remember his high boots and the wonderful jackets he wore. Prince Juan Carlos of Spain was there for the races too, and King Olaf of Norway, who's a wonderful man.

"I guess you could say it was exciting."

"He's put me in a role. I have to accept it to be with him. Silence is my best weapon. If I say something, he will just turn it around. He's a great debater. So I'm often silent.

"It's difficult, but I'd say I'm happy. I just sit back – I enjoy the show."

BUYS FIVE RADIO STATIONS

Jim Roddy: (When Ted buys worst radio station in Chattanooga) "There was a teenager with a nasal condition who played rock albums from his own collection at home.

"The manager loved golf. You ever hear of a radio station covering golf? This station, I mean, they covered golf. Live. On radio. So we changed right away to a Top 40 format, and brought in young announcers with good voices, and it was all right."

BILLBOARDS AND RADIO – A HAPPY MARRIAGE

"See, I could take my vacant signs and promote my radio stations. One of the things that was wrong with the billboard business was that although we had a 25 percent profit, we would also have 25 percent of our signs not being used. It seemed like an awful waste. So I would put up my own radio ads on the open billboards, and use them that way.

"It doesn't sound too brilliant, but it worked. I got the idea from an oil refinery. It had an open flame on top. The heat was being lost uselessly, and I wondered why they didn't put it under the cooker and make oil with it.

"It's fire, right? At least put it under a coffee pot. That bothered the hell out of me, believe it or not – that little wasted flame. You really have to be dumb to waste your resources, because you've only got so much."

BORED WITH RADIO, TED SEES TV AS THE BIG TIME

"When I bought Channel 17, everybody just hooted at me. The station was really at death's door.

"I didn't bull shit anybody. I told them I didn't know anything about TV. I just love it when people say that I can't do something. There's nothing that makes me feel better, because all my life people have said I wasn't going to make it.

"The secret of my success is this – every time I tried to go as far as I could. When I climbed the hills, I saw the mountains.

"Then I started climbing the mountains..."

EYEWITNESS

Irwin Mazo: "Gawd almighty! The station was within thirty days of going under. Ted got the idea that he wanted to buy it. Well, I had been through one big crisis when he took the billboard company back from Bob Naegele, and I said I just couldn't take it another time. Jim Roddy was with me. He said, 'Why are we doing this?' It was just an atrocious business proposition." (Irwin Mazo quits)

EYEWITNESS

Lee McClurklin, Robinson Humphrey: "The guy's really got balls. If you were given the opportunity to mortgage your home with a 25% interest second mortgage to make a payroll offer on a television station you didn't need, would you do it?"

ATLANTA TV STATION IS NUMBER 5 out of 5

"I had never watched the station, because I couldn't even get it on my set. I never watched any TV in those days. I had no idea what UHF stood for."

THE STAFF?

"Mainly hippies and inexperienced college-kid type engineers. And the whole operation was terribly undercapitalized."

"To me the whole damn thing was a challenge...

"One of the first things that I learned was that outside of New York and Los Angeles, there's no way two indies in the same market can make money. Especially two UHF's.

"I knew that one of us would be off the air before too long. And I was going to be damn certain it wasn't us."

SELLING ADVERTISING BY CLAIMING SMART VIEWERS

"What makes them so smart? Because you have got to be smart to figure out how to tune in a UHF antenna in the first place. Dumb guys can't do it. Can you get Channel 17? No? Well, neither can I. We aren't smart enough. But my viewers are."

"I felt the people of Atlanta were entitled to something different from a whole lot of police and crime shows with murders and rapes going on all over the place.

"I believed – and I still believe that people are tired of violence and psychological problems and all the negative things they see on TV every night. I wanted to put on something different and give them a choice. Also I wanted to prove that a small, locally owned, independent, UHF station could make it in a big market."

EYEWITNESS

R.T. Williams: "The joke was we were number seven in a five station market. There was another Atlanta independent (UHF) and they were just killing us.

"But when their parent company pulled the plug they suddenly went dark. Bang! And I got a call from Ted who wants to do a 'Thank you, Atlanta' party. A one hour special program. I hang up and go roaring into his office and tell him, 'Hey, Ted, we didn't win! We didn't beat 'em! They defected!' He says, 'Nah, we beat 'em. We won.'

"As sure as shootin', we put on a one hour special. We had a couple of bands, with balloons floating down... and Ted was wandering around talking and thanking the people of Atlanta for their support. I'm thinking, what is this bull shit? I've gotta get out of here!

"I look back on it now and remember that nobody knew who the hell Ted Turner was at the time, but he CALLED it a victory and so it WAS one."

TO STAFF IF THEY LOST AN ACCOUNT

"Never look back. Don't worry about it. Just keep moving forward."

RECRUITING STAFF

Interview question: "Are you a dreamer?"

EYEWITNESS

Gerry Hogan: "In our first meeting, he was amazing. He was like a hyper-active kid on Christmas Eve. He was running around the room – and he'd walk behind me and spin me around in the chair.

"I said to myself, well, what have I got to lose? I was twenty-five and I figured it was at least worth investing a year to see what happens.

"Every year from then on, I said the same thing. It's like you're half way up the hill and you've got to see what's on the other side. Every year something crazy or wild came up. Eventually you take on an attitude that you're like guerilla fighters. Or like a bunch of pirates going after an armada. You accept the tough parts, because there's a sense of mission. I got to the point where I actually enjoyed living on the edge, which is what Ted always did. And he would do anything, like stand on a table or take a guy by the throat or kiss his feet, whatever was called for in the situation.''

HIRING JIM RODDY

"You've got to come. You've got to come. I don't know where I'm going, Jim, but hang on to my coattails and we'll get to the stars and the moon."

EYEWITNESS

Jim Roddy: "It was a complete disaster...

"The first thing I learned about Ted was that he was very, very, bright in addition to being completely wacko. I'm more conservative. I like to have the bills paid, money in the bank, and the planning done for a year. But Ted can't stand a nest egg; he's been a plunger from the beginning...

"Ted, as an idea man, is one of the most innovative people alive. But he is not an administrator at all. He is a free-thinker with extraordinary business judgement – but off we went, anyhow."

EYEWITNESS

Gene Wright: "There were maybe forty-three people in the whole corporation when I arrived. We had absolutely no equipment at Channel 17, and no real engineering or maintenance being done. We had all kinds of technical problems and kept going off the air.

"I bought a cot and basically moved in and stayed there. The operators would sit around playing banjos and smoking pot while the film would run out...

RADIO AND TV – TALKING BILLBOARDS

"We had a bad transmission line... Every time the wind blew and the tower moved our transmitter would cut off. Ted wouldn't buy a new line.

"Finally, when we were off the air several more times, he asked what another one would cost. I said it would be around $75 thousand.

"By the time he agreed, we had waited too long and our transmission line blew all the way up at the top. It took about seven days to get back on the air. I worked up on the tower and never got any sleep and my shoes were full of blood. If I had time to look for another job, I wouldn't have been there.

"I thought he was crazy.

"I'd say, 'We need another monitor.' Ted would say, 'Well, you've got one.' I kept insisting that we needed 'redundancy' in the system. He said, 'Gene, you are the most REDUNDANT BASTARD I ever met!'

"I stayed around because of the challenge. After about three years of working seven days a week, with no vacation, we started making progress. Ted came through one day and he said, 'Hey, Wright, we haven't been off the air for a long time, have we?'

"I said, 'No, we haven't.' He said, 'God, redundancy's great!'"

ON FIXING CHAIR WITH BROKEN LEG IN HIS OFFICE

"No, no, I don't want to do that. When people come to sell me stuff, I don't want to look like we're doing well. I want them to feel sorry for us, have them feel they're helping me out by giving me a good price."

BUYS WORST TELEVISION STATION IN CHARLOTTE

"The first year I owned Channel 17, we lost over five hundred thousand dollars. Wow! But at the same time I found another station that I could get cheap - a UHF in Charlotte, North Carolina.

"I told my board of directors that I wanted that, too. They said, 'You gotta be kidding!' So I said, 'O.K., I'll buy it myself, and I did. With my own money.'

"Irwin and the others thought I was crazy, and in all fairness it looked pretty dark there for a while."

EYEWITNESS

Bob Schussler: "When I got to Charlotte to head up sales, I wondered what the hell I'd done. I mean, it was out in the country. We had woods on both sides and, in front, a pasture with cows.

"The second night I was there, I was watching our big movie in prime time. Our highest rating, by the way, was a two, which is about as low as you can get. So I'm looking at the feature which starts at eight. It's some Maureen O'Hara movie about the desert. Here it comes on and you see these camels with Maureen O'Hara – and they are walking upside down and backwards! This goes on for about three minutes and they shut it off.

"It's just black. For five more minutes. Then the movie comes on again – but the camels are still upside down, so it goes off again. But this time instead of a black screen they run about seven minutes of straight commercials! Finally, it comes on, and we try once more – but now it's a different movie!"

A BEGATHON - TED ASKS VIEWERS TO SEND HIM THE PRICE OF TWO MOVIE TICKETS

"I had to actually go on the air in Charlotte and solicit loans from the viewers. I said it's a telethon – a telethon for me, because I need the money to get the wolf away from the door. The wolf was breathing all over me. The telethon took in $35,000, just as gifts to us.

"But do you know what?

"I paid all those people back within three years. If I ever need money in the future – I know how to get it. I'll just go on TV and say, 'I'm Ted Turner and please send me some money. I promise to pay it back with interest.'"

PROGRAMMING CHANNEL 17

"I programed the whole station myself in those days. I sold ads, I signed all the payroll checks, I went to parties, I met people, I asked a million stupid questions, and I educated myself. I would wander around in a daze all day thinking, 'What am I going to put on at four-thirty? No committees, no studies, no bull. I would ask Janie what she thought of a certain movie, or I'd ask one of my friends, or the girls at the office. I started spending a lot of time with the film people – MCA, Paramount, United Artists, Viacom – all of them. And they helped me whenever they could."

DRIVING AROUND IN A DAZE

Press: "Turner seldom negotiates the thirty-minute drive into downtown Atlanta from his modest, magnolia-trimmed home near suburban Marietta without getting hopelessly lost. He has been making this trip for four years. 'Someday I know, I'll end up in Tennessee,' he says."

SAILING WAS STILL TURNER'S FIRST LOVE

"Atlanta's great in the Spring and Fall. In the Summer, I'm off sailing, the Winter is broken up nicely with the SORC and Montego Bay, and then I come back in the Spring. It works out nicely."

REESE SCHONFELD OFFERS NEWS PACKAGE TO WTBS

Sid Pike (station manager): "NEWS? What do I want NEWS for?

"We don't need news! We counter-program against news. News loses money! We MAKE money. We don't do news.

"I work for a crazy man and he buys so much entertainment that – oh hell, we've got more movies here than we can run! I can't even play the stuff he buys NOW – much less put on a half hour of news! We'll never do news."

ON NEWS PROGRAMMING

"We didn't do the news seriously on my station because we just didn't have the budget to do it properly. My father always told me, 'If you can't do it first class, don't do it at all.' It wasn't that we didn't want to do it. We couldn't do it properly so we didn't do it at all. There's nothing worse than looking silly, you know. In those days I used to get kidded by my friends in the television business who were saying, 'You might be doing O.K. in the ratings, but you haven't got any news.'

"And I would say, 'Well, you just wait. One of these days, I'm going to come on with news that'll make y'all green with envy.'"

CHAPTER FIVE
THE BRAVES, THE HAWKS, THE FLAMES, THE CHIEFS

"I had this horrible, recurring, nightmare in which I picked up a newspaper and the headline said:

'BRAVES MOVE TO TORONTO.'

"I was out at the ball park rooting like always and there were maybe two hundred people in Fulton Stadium. The team really stank, ugh.

"I went up to see Dan Donahue, who was the team's president, and I caught him peeking out of the glass in the owner's box.

"I'd had a couple of beers and I was really worried about selling ads for the team the next season, so I said, 'Hey, Dan. What are we going to do about getting these Braves going?'

"He said, 'I don't know what you're going to do, but we are bailing.' 'Oh, my God,' I said – and almost had a heart attack. 'Who're you going to sell it to? I've got a five year contract and this is only year two.'

"He said, 'To you.'

"'To me? For how much?' 'Oh, about ten million.' And I said, 'Yeah? Well – how much is it losing this year?' 'Oh, about a million bucks this year.' I couldn't believe it! 'What, I'm going to pay you that much so I can lose a million a year, too?'

"These guys gotta be crazy, I thought."

"It turned out that the Braves had one million dollars in their own organization, so I bought it using its own money which was quite a trick."

ON BEING TOLD HE'S A GENIUS AFTER THE BRAVES DEAL

"A genius? Me? That's really baloney. Who told you that? Albert Einstein is a genius, not me. I know exactly how smart I am. My IQ is 128. That ain't bad, but it isn't the point. It doesn't help you make a good business deal – that I learned from my father. That, and work your butt off, but not from behind a desk like most people do."

JUST BEFORE BUYING BRAVES

"What do you need to know about baseball?

Both sides have ten guys."

"We televise all the Braves' away games, and we got the Hawks away from ABC. Big bucks. It costs us money to televise the Hawks, but it's bringing in viewers. Big bucks. But a faint heart never won a fair maiden."

CONVINCING BOARD OF DIRECTORS TO BUY BRAVES

"This is something we HAVE to do. We need to do it. If we don't do it we might lose the rights – Who knows? And then how are we ever going to be a major player in the business?" (Won approval)

THE BRAVES, THE HAWKS, THE FLAMES, THE CHIEFS

AMIDST POPPING FLASH BULBS AND PRESS

"We're going to operate freely and openly. Atlanta is my home and I love it here. I've been all over the world racing and stopped in Tahiti on my way back from Australia which I read a lot about as a little boy, and I decided the closest thing to Paradise on this Earth is Atlanta, Georgia.

"We made some money in the outdoor advertising business here over the past ten years and invested those profits in Channel 17, and I think our involvement has gotten deeper and deeper as far as Atlanta's sports teams are concerned, and I think we're interested in all the Atlanta sports teams doing well and...

"I want to see nothing better than the championships for all four major sports come to Atlanta.

"And, since the television station has done pretty well and become semi-successful, we're taking our profits from that and reinvesting them in the team.

"...I think all of us know, it's going to be a big job bringing the World Series here to Atlanta, but that's going to be our objective, hopefully within five years. It takes time, effort and money, but that's something we're going to dedicate ourselves to!

"I don't want to see any more headlines in the 'Atlanta Journal-Constitution,' bless their souls, that call Atlanta 'Losersville, U.S.A.' I want to see - 'Winnersville.' With a little luck and the Lord's help – that will be our short and long term objective.

"I really don't know as much about baseball as I should – I'll be quite honest with you. But I intend to learn, and learn as fast as I can with everyone's help.

"Turner Communications is buying one hundred percent of the Braves, but on an installment basis. We're going to owe the Atlanta LaSalle Corporation quite a bit of money.

(How much did you pay?) "The price is undisclosed. The lease on the stadium is something we do inherit, and it has about fifteen years left to run.

"The buck does stop at my desk, and if anybody wants to take a swing at someone I guess I'm the one to take it at, and – I hope Cassius Clay is not listening... but I'm sure I could give any of you all a good fight... a middle-aged media man like myself!

"It's not an economically wise move to buy the team but money is not the prime motivation here. I believe I'm doing it primarily for the city and the southern part of the country, believe it or not.

"The little nuances are what make the difference between winning and losing...

"I'm thirty-seven now, and the team won't be sold again in my lifetime."

THE TED TURNER BRAVES 400 CLUB SPEECH

(At podium in Hank Aaron tie): "... I'm sick of mottos. ... If we have one at all it's going to be 'VICTORY OR DEATH.'...

"If things get bad enough, and they may, we'll lock up the stadium, play day games to save electricity, and, by God, if anyone asks what the Hell we're doing in there, I'll tell them 'WE'RE LEARNING HOW TO WIN!'...

(Lights cigar - for third time) "I might even sign up John Zook of the Falcons and use him as a pinch runner. We'll kick the Hell out of somebody!...

"I've got some friends in the Mafioso up north and, as much as I'd hate to resort to these kinds of tactics, we'll rough up Pete Rose or Dave Concepcion if they start playing too well. (By now the dinner guests are responding with a mixture of shock, hilarity and absolute fear.)

THE BRAVES, THE HAWKS, THE FLAMES, THE CHIEFS

"... And I'm seriously thinking of changing the name of the Braves to the Atlanta Eagles – it's cheaper to feed a bird than an Indian. Whatever we do, we're going to try to make baseball fun again for you. (Huge applause)

"We may lose it all on this deal... but if everything goes down the commode... they'll have to come get me – and Channel 17 too!...

(After keeping guests in laughing disbelief for half an hour) "They told me I had five minutes up here to talk to you. I guess I got kind of carried away."

TURNER'S LIST

"1) Make baseball fun again at the stadium.

2) Get Cracker Jacks back in the stands."

NEWSPAPER HEADLINES "TURNER, ATLANTA, TOGETHER IN COMMON LOVE"

"I never could understand why owners like to sit up behind bullet-proof glass sipping martinis. I sit in the front row."

"I bought the Braves because I'm tired of seeing them kicked around.

"I'm the little guy's hero. They love me. I run the team the way they think they would if they owned it. I come to all the games. Sit in the stands. Drink a few beers. Even take my shirt off. I'm Everyman to them – their pal Ted."

In front row behind dugout: "Awwriiight!" (Moans) "One and nine."

"It takes the edge off life. Maybe I should have gone to see my twelve year old play Little League instead. My wife's thinking about suing the team for alienation of affection."

(Ball flies overhead) "There goes four dollars. (Three more foul balls follow) SIXTEEN dollars!"

TO CHIEF NOK-A-HOMA

"Chief, we gotta grow a whole lotta corn this summer, 'cause we sure can't live off the baseball attendance."

TO LITTLE OLD LADY

"Pleasure havin' you here, sweet thing."

TO GUM-POPPING CHEERLEADER

"Darleeene... Are you chewing tobacco?"

TO BLACK POLITICIAN

(Who had just paid Ted a compliment) "David, you're going to end up the first black in the White House."

CONDUCTING CROWD IN SINGING

"Pack up your troubles in your old kit bag and smile, smile, smile!"

WHEN BRAVES WERE LOSING

(Grabs mike in press box): "This is a new policy for the time being. Nobody's going to leave here a loser tonight. If the Braves don't win this one tonight, I want you all to come back tomorrow night as my guests. I'm in this for life and we appreciate your support. We do. We're going to beat Hell out of the guys who're beating Hell out of us right now!"

"All the coaches were chewing and somebody offered me a chew. They were teasing me, so I took it and chewed it. They all stood around waiting for me to get sick, but I fooled them. I liked it."

"What the Hell, I love these guys. The old baseball 'muckety-mucks' don't know what it's like to get to know their boys. These guys are the finest bunch of fellas I've ever been associated with and I love them."

TO TERRY MCGUIRK, WTBS SALES STAFF

"Get ready to leave tomorrow morning. We're going down to Florida for Spring training."

WITH MCGUIRK IN THE DUGOUT

"Why did you run to second base?

...What the hell is a balk?"

After spring training: "It's a lot tougher out there than I really thought. That damn ball comes at you one hundred miles an hour. I'll be a batboy some night, and I'll sweep the bases, too. I might even sell concessions. That'd get 'em!... I might even pitch an inning, too!"

EYEWITNESS

Terry McGuirk: "He had never done anything with a ball when he was growing up. He had no hand-eye coordination, and he really didn't know what was going on, but by watching me and asking questions he could get a picture of how easy or difficult it might be. We had a lot of fun – and by the end of spring training he had a good handle on the game of baseball, before he went on to do many other crazy things."

EYEWITNESS

Harry Reasoner: "I left that baseball field liking Ted Turner. I'd hate to have to keep up with him, and I'd hate to be that driven myself. But I liked him. He's a talented man with a very large ego, not unheard of in American business and show business. And when you look at his accomplishments, maybe you can't blame him. He's just convinced he can do almost anything he sets out to do. And so far he's been right."

TED'S LIFE "BECAME" THE BRAVES

"That first year I was really active. I devoted half my time to the Braves – which is a huge amount. I went around to every National League ball park, met all the general managers, discussed ticket pricing, how to develop players, even how you print up a program. I didn't have any idea how complicated it all was...."

"It was a nightmare."

"The first thing I did was spend a million dollars on a giant TV screen over the scoreboard."

EYEWITNESS

Hammerin' Hank Aaron: "Ted has worked very hard since I've been here. I was amazed at the amount of knowledge he'd picked up in such a short time...

"I am the vice-president in charge of player personnel, and that means, of course, the minor leagues. I am over the minor league system.

"I'm just happy that Ted saw to it to do some of the things that other owners had been afraid to do – like elevate blacks. Bill Lucas and I have responsible jobs in the organization, and even if we don't stay here but a year, I've got to feel that he opened the doors not only showing people that blacks can operate but, just given a chance, they can do almost anything they set their minds to, and I'm very grateful to Ted for this, too."

ON REVERSING THE FORTUNES OF THE BRAVES

TURNER: "I'll do it because a lot of people in high places laughed at me. Watch me. I'm like a bulldog that won't let go... Why do you think my own racing yacht is named 'Tenacious?'"

PLAYBOY: "We give up. Why?"

TURNER: "Because I never quit.

"I've got a bunch of flags on my boat, but there ain't no WHITE flags. I don't surrender. That's the story of my life. Just think, if you were a rabbit, to survive, you'd have to hop fast and keep your eyes open:

"Ride, boldly ride, the shade replied – if you seek for Eldorado."

TED SEES PROFESSIONAL SPORTS AS GAMES

"Life itself is a game,
and money is how you keep score."

"I'd rather sink than lose."

"If it works it must be right."

"The fans want to see baseball
– and we're going to give them baseball."

LEARNING THE LINGO

"Not too shabby!..... Throw strikes.... watch for the slider... Take 'em downtown!"

ATTENDANCE TRIPLES WITH TURNER'S ANTICS

The nose ball race: "Oh God, why do I do this type of thing, man? I'm going to end up killing myself someday. (After scraping his nose and face and bloodying the ball)... I beat Tug by a mile!...

"When you're little you have to do crazy things. You just can't copy the big guys. To succeed you have to be innovative."

IN SILK JOCKEY OUTFIT (NUMBER 17) TURNER LOSES OSTRICH RACE

Frank Hyland, publisher of 'Atlanta Journal'(and winner of the ostrich race):

"I never even knew her name. We had been together, man and beast, for a brief period of time, for that one moment of truth, working and driving together for the roses. I shall never forget her. And then the van of ostriches took off through the night. The birds tried to kick in the sides...

"A tear came to the eye. I never even knew her name. Or was it a him?"

TRADING PLAYERS

Cajoling Andy Messersmith, the "million dollar baby": "All Andy wanted was a no-trade clause in his contract so that he wouldn't wind up in a Detroit or a Cleveland. As good as HE is, the Dodgers should have given it to him. I kind of feel like we cheated, like we stole him. But all's fair in love and baseball."

"He'll never be traded. He'll be a Brave as long as I am...

"It's for life or until death or old age do us part."

GOING AFTER GARY MATTHEWS

To owner of Giants: "Anything you offer Gary Matthews, I'll offer him more."

(Inviting owner of Giants to party in Atlanta for Gary Matthews) "Everybody is going to be there to welcome Matthews to Atlanta; the governor, the mayor, everybody...

"There's no law against showing a player our fine city. You know I'm not going to make him an offer. That would be illegal, at least before the re-entry draft deadline – I just want Gary to be aware of what a wonderful city he can play in. Then when the time comes I'm going to offer him more money than the Giants will take in next year.

"Of course, it may take much, much more than that."

Circulating at the party: "Go shake hands with Gary... Tell him what a great city this is."

"If it were me, I'd be moved by this kind of turnout. I just don't believe money is everything to these guys. Anybody wants to feel wanted."

Gary Matthews: "This is nice, real nice."

GARY MATTHEWS JOINS BRAVES

Matthews: "Ted could talk you into anything. I turned down two million dollars to play here - and Ted is the reason..."

ON BASEBALL AGENTS

"When they smile blood drips off their teeth."

BRAVES IN LAST PLACE FOUR SEASONS STRAIGHT

"It's good for my humility. As my father used to say, 'A few fleas on a dog remind him what he is.'"

BASEBALL COMMISSIONER THREATENS TO SUSPEND TURNER FROM BASEBALL

BOWIE KUHN: "Why can't you be like everybody else?"

TURNER: "Because I'm in last place."

"A baseball owner's got less rights than any regular US citizen. It's like all of a sudden being shipped off to Russia. I may be kind of weird, but is it against the law to have a good time?"

HOW THE LITTLE GUY CAN BEAT THE BIG GUY

"If you want to get to the top you've got to argue with the top. If there's a big guy and a little guy in an argument, if the big guy will argue with him, the big guy doesn't come down to his level. The little guy rises up to HIS level. Now I'm in a fight with Bowie Kuhn. He's big and he's important, and he's a commissioner. I'm going to fight him for all he's worth so long as he'll fight back so I can rise up to his level.

TO BASEBALL COMMISSIONER

"Great White Father, please tell me how to avoid fighting for what little we have left. The buffalo are gone. The white man came and killed off all the buffalo.

"They drove the trains through what we were told we would have – this land, you know? The Black Hills. Now this gold you want – the yellow metal – you want us to leave and go to the dust bowl of Oklahoma and these are our homes. We must fight for them.

"Please go back to the Great White Father, soldier man, and tell him to please help us...

"I am very contrite. I am very humble. I would get down on the floor and let you jump up and down on me if it would help. I would let you hit me three times in the face...

"I would bend over and let you paddle my behind, hit me over the head with a Fresca bottle... something like that. Physical pain I can stand."

To Bob Hope, Braves' Promotional Manager: "Well, there are a couple of things (the commissioner) could do. First of all, he could return Gary Matthews to the Giants. But that wouldn't be punishing me. That would be punishing him. Or he could fine me a lot of money. But I've got a lot of money. If he fines me a lot of money, he knows I pay it and then I go on my merry way. Or he could REALLY punish me, and at the same time get me out of his hair, by suspending me from baseball. When we go to the winter meetings, we've got to make sure he suspends me from baseball."

TURNER SUSPENDED FROM BASEBALL

To press: "Just say that for once in his life Ted Turner was speechless – that he was at a loss for words."

"The world has gotten along without Abraham Lincoln and John Kennedy... Baseball can get along without me for a year."

SUES BASEBALL COMMISSIONER

(After cross-examination in court): "After this is over you keep that up and you're going to get a knuckle sandwich."

"The man questioned my honor."

QUOTING NICK FROM 'THE GREAT GATSBY'

"Everyone suspects himself of at least one of the cardinal virtues, and this is mine; I am one of the few honest people I have ever known."

IF YOU BUILD IT, THEY WILL COME

"I started off saying we'd have a World Series in Atlanta within five years, but I've had to move the date up a little. 'We'll prove it to you in 1982,' yeah. I just have to have time to learn the hard way, like always."

1991 BRAVES BRING WORLD SERIES TO ATLANTA

Eyewitness: "In a closely contested battle with the Minnesota Twins, three of the seven games went into extra innings, four were decided on the last pitch and five were won in a team's last at-bat. The seventh and deciding game was tied in the tenth inning, with the Twins finally winning 1-0 to take the series.

"As one of his Atlanta Braves headed toward home plate after hitting the only grand slam home run of the series, the Superstation's national coverage cut to Ted Turner, fast asleep in the stands." (In 1995, the Braves finally won the World Series.)

LOOKING BACK

"Before I got into baseball, hardly anyone knew who I was."

THE OLD BALL GAME

"Some things don't need changing.
The sunrise doesn't need changing.
Moonlight doesn't need changing.
Azaleas don't need changing.
And baseball doesn't need changing."

KNOWING CHANNEL 17 WOULD SOON BE UP ON THE SATELLITE, TURNER BUYS 55% of ATLANTA HAWKS BASKETBALL TEAM FOR $4 MILLION

"I own the worst baseball team and the worst basketball team, I must be crazy."

GO HAWKS!

"Outnumbered five to one, there they were. Alexander the Great digging in with his famous phalanxes. Nine men deep, with swords of different lengths over the shoulders of the men in front. Oh, maybe it was five men deep. God, how could a guy carry a sword that long? Alexander the Great shocked them. Just like the Hawks. Goddamn. Go HAWKS!"

EYEWITNESS

Pat Williams, manager of the Philadelphia 76ers: "Ted Turner is every kid who ever got loose in Disneyworld."

NEXT SIGNS ATLANTA FLAMES HOCKEY, AND IS BLOCKED FROM BUYING ATLANTA CHIEFS SOCCER

"Yeah, but I wanted soccer for my TV station, so I loaned every penny of the purchase price to a friend of mine, so he could buy them. I mean, I loaned him every dime. There was nothing in there said you couldn't lend money to other owners. The league knows. They said, 'Oh you LOANED the money. Well, we can't do anything about that.'

Ha! See, they can't stop me. I'll be like the Man Without a Country one of these days. He went on living, but just from different ships."

"The Braves, the Hawks and the Chiefs – they'll have fans all over the country. You should hear the heated debates at league meetings – people almost screaming at each other. I say, 'Gentlemen, the law of the land takes precedence over the law of the league.'"

"By the time the leagues found out what I was doing:

"The horse was out of the barn.
Yeah, the ketchup was out of the bottle.
The plane was out of the hanger.
The toothpaste was out of the tuuuuuuuube."

CHAPTER SIX
CAPTAIN COURAGEOUS – UH – OUTRAGEOUS

AGE 11, "THE CAPSIZE KID" LEARNS TO SAIL AT SAVANNAH BEACH YACHT CLUB

"I didn't have the ability to play baseball. Couldn't swim – almost drowned. I tried track – ran the hundred yard dash in fifteen seconds. But I knew I could sail. I just kept working and working."

EYEWITNESS

Florence Turner, Ted's mother: "Ever since he was a little boy, Ted's always been trying to win. When he was first sailing Penguins the wind would come up, and all the other kids would get their sails down and wait for a tow. Not Ted. He would just keep going until he turned over...

"He doesn't get it from me. I never cared if I won or not."

GRADUATION GIFT FROM McCALLIE, "THE GREASED LIGHTNING"

(On Lake Erie) "I was so impressed just to be there and meet all the real big cheeses."

(After winning sixth place in Lightning class) "Look at the best sailors in the world! Here we are! Coming in sixth place!"

ON WIND CHANGE

"You can see it. Look out there in the atmosphere. Then use your nose. You can smell where the wind is."

OCEAN RACING ON NEW YACHT, "VAMP-X", NAMED FOR HARD-HEARTED HANNAH, THE VAMP FROM SAVANNAH

"Some races are so beautiful you're sorry to finish. There are times when the weather is perfect. The Montego Bay race in '66 was one. It was a full moon, clear nights, warm, and there was a good wind. The world was BEAUTIFUL. In our sport, you're out there with nature – you're as close to nature as you can be – with gulls, flying fish, whales, the dawn, the sunset, the stars. You take a deep breath – and you feel alive, really alive. The brilliance of the stars is hard to describe. You think you can reach out and grab a handful of them. It's as if they are ten feet away, and there are millions of them.

"There are so many beautiful sights.

"The coast of Tasmania with its cliffs that go 1000 feet straight up. Sailing past Molokai, in Hawaii, in the moonlight after not having seen land for two weeks; storming past the coast of Cuba; seeing Bermuda loom some 70 or 80 miles away, or seeing the phosphorescence in the tropics. It's just so BEAUTIFUL out there."

WITNESSES AN AMERICA'S CUP RACE WHILE AT BROWN

"We were on a sailboat about thirty feet long owned by the family of a friend of mine. We were near Castle Hill, and they towed the boats by – both white, if I remember. The crews were all big, muscular men, with their matching shirts on. I'm sure that at the time I didn't just decide I was going out and win the Cup, but I was pretty impressed."

CAPTAIN COURAGEOUS – UH – OUTRAGEOUS

PARAPHRASING 'NIGGER OF THE NARCISSUS' AND 'YOUTH' AFTER A SYDNEY-HOBART RACE IN HIS NEW YACHT "AMERICAN EAGLE"

"Ah! The good old time – the good old time. Youth and the sea. Glamour and the sea! The good, strong sea, the salt, bitter sea, that could whisper to you and roar at you and knock your breath out of you.

"The crew of the 'American Eagle' drifted out of sight, I never saw them again. The sea took some, the steamers took others, the graveyards of the earth will account for the rest.

"So be it! Let the earth and sea each have its own."

WITH PHOTO ALBUM

"Now look at this. This is me in 1966 at the SORC races. That's 12 years ago. Super! I sort of look like Errol Flynn there. What the hell, I haven't changed any. Except I was brash and loud then. Here's the reason the New York Yacht Club hated me."

CHOSEN AS SKIPPER FOR "MARINER" IN 1974 AMERICA'S CUP

"If the South had won the Civil War, I would have been a Confederate challenger, not a U.S. defender."

THE SPEECH THAT GOT TED INTO THE NEW YORK YACHT CLUB

Bob Bavier: "Look, you want to know the faults about this guy? Ask me. I can give you a bunch as long as my arm. But you won't find one you haven't already heard about. All his faults he wears right on his sleeve. But I know this fellow pretty well, and he's a straight shooter. I like him."

TED'S FATHER WOULD HAVE BEEN SO PROUD

"I remember him telling me about the New York Yacht Club once. How it was swank and ritzy and all. My father never would have dreamed of my being in that room. That I'd be a MEMBER. That I'd defend the America's Cup. Lord, it would blow his mind."

NEWPORT IS KNOWN FOR ITS STATELY "COTTAGES", GREEN HEDGES, AND WELL-BRED ETIQUETTE

"You're not sailing your own boat. You're working for a committee, the New York Yacht Club Selection Committee, and you're working for a syndicate. So you have to be a little bit of a politician, in the better sense of the word. You have to be a gentleman, and you have to do what is expected of you on the water as well as off. I mean, going around and writing 'turkey' with a grease pencil on another guy's boat, like you do in the Finn class, doesn't make sense in a twelve (meter). If they found out who did it, you'd be taking a walk down the dock the next day." (So Ted harnessed his freewheeling style, as he said, within "chains of gold.")

"MARINER" HAD A DESIGN FLAW WITH A SQUARED-OFF STERN, AND KEPT LOSING

"You can always change skippers if they are hacking things up. You can even change the cut of your sails, as 'Courageous' did in July while fighting for the right to defend the cup. But you can't change the design of the boat."

ON LOSING RACES

"I'm like the grass. I get trampled down one day and spring right back up the next. I've been beaten so much that one more loss doesn't make any difference. Losing is simply learning how to win."

AMERICA'S CUP TRIALS IN NEWPORT, 1977

TURNER: (Looking down over two huge white beauties, "Courageous" and "Enterprise") "You know, when you were fifteen years old, did you ever think you'd be sailing on a twelve meter?"

GARY JOBSON, tactician: "I never would have dreamed it."

TURNER: "Neither would I. Isn't that the greatest thing that ever happened to you?"

EYEWITNESS

Gary Jobson: "He'd get us all together in the morning, start waving his arms, and then he'd say, 'Hey, isn't this a great day to be alive? Today, we've got the opportunity of a lifetime, the chance to go head-to-head against the world's best sailors. Let's go have some fun.'

"My deal on the boat, my strength, was to never allow myself to be surprised. I was always ready with good advice based on plenty of preparation. Ashore our crew was self-motivated. Turner had selected well. That may be his greatest talent, in fact. He is best at selecting people and telling them what to do. But the crew members had to get their jobs done in spite of Turner. When he left them alone they were fine. But once in a while he would come around and start telling them how to do it...

"So I would grab him and suggest a walk, an ice cream cone, or remind him that the Braves were on TV – anything to get him off the boat. My job was keeping Ted from going off the deep end...

"I do know that Turner became the real challenge, not the Cup."

THE TRIALS BEGIN

"I mean, there was no way we could lose. We wanted it so much after 1974. We were the best sailboat crew in the history of the planet, and we knew it."

"I can make eleven guys work harder and longer than anyone else."

"We came to win and that's what we're doing, right?"

BERTHS ENTIRE BRAVES TEAM ABOARD HIS OWN YACHT, "TENACIOUS"

"If I have to watch them lose all the time, they might as well watch me try to win."

EXPLAINING SAILING TO THE BRAVES

"Tacks are like snow flakes. They all look alike but every one is different."

MILITARY TALK

"I really think that man can get through life without violence to the death. That's why I like sailing. You can have all the thrill of killing somebody without killing them."

"The most fun that you ever have as a man is in doing men's things. Men's things are primarily getting a bunch of guys together and going out and conquering a country, fighting a war, winning a big fight, putting a baseball team together. But first of all, you've got to get a good bunch of guys together and do it, whatever it is. And then you have to get them all excited and motivated so they'll bust their ass."

FINAL RACE IN JULY TRIALS

"We've got 'em! It's all over baby! We won! Thatta way to go, Robbie. That new jib sure helped. AWWRIIGHT! AWWRIIGHT!"

"I'm going to do something that nobody's done before. I'm going to enjoy the final trials. Win, lose or draw."

AUGUST TRIALS RACING TO THEME SONG FROM 'ROCKY'

(To crew) "We need to win because we're just a bunch of bums. Are there any bluebloods here? Raise your hands. None. I didn't think so. Anybody from fancy prep schools? Awwriight. We could have won it with the latest in computers, but that's not our style. We like to slug it out. These other guys are going to have to slug it out with a bunch of street fighters from the Bronx and Queens."

"We didn't use tank tests or anything. We didn't even have an onboard computer – except for a thing we got near the very end just to keep track of where we were in case it got foggy, which it never did."

AFTER BEING KICKED OUT OF HALF THE PUBS IN NEWPORT FOR "OFF-THE-WATER INDISCRETIONS"

"If being against stuffiness and pompousness and bigotry is bad behavior, then I plead guilty."

BARRED FROM THE BLACK PEARL

"The guy was acting like the king of a mountain, crowing about his money and being a pompous ass. Look, I had to put him down. I apologized later in the summer, but I just can't stand phony airs. That's the whole thing. The phoniness in this world. All this just ticks me off. Injustice. I hate injustice.

"I have such a distaste for people who can't roll up their sleeves and get the job done. See, I'm not wrong. I'm just getting a bum rap. I'm not a wild man. I'm not a bad guy.

"I'm not like some (ball club) owners I could name. Pro sports has become filthy; it's a double-dealing, rotten business. There's all this terror and intrigue and fratricide. I'm sick of it. Sports should be fun. My Braves and my Hawks play clean. They may be underpaid but they have fun. They're HONORABLE. In yachting, men understand that. You're assumed to be a gentleman. You are assumed to have honor."

NATIONAL PRESS COVERAGE

Time Magazine: "During the Cup eliminations, he flirted with every girl in sight, crawled pubs with his crew, got tossed out of chic clubs and restaurants for boozy behavior, and turned Newport's bluebloods positively purple."

CAPTAIN COURAGEOUS ON THE BLUEBLOODS

"I just can't stand snobbery or phoniness. I mean, I just can't stand it. It makes me ill."

EYEWITNESS

New York Yacht Club member: "If he's ever selected, he will be the first skipper in Cup history to appear at the starting line wearing a muzzle."

ON WINNING FINAL AUGUST TRIAL RACES

"There will never be a time in my life as good as this time. I can't believe this is really happening to me.

"I'm so hot I just tell my guys to stand by me with their umbrellas turned upside down to catch the stuff that falls off me and onto them.

"And I can't think of anything I could do to be any better. My biggest problem now is to keep from getting a big head. You guys don't think I have one, do you?"

"COURAGEOUS" TO DEFEND THE UNITED STATES

NY Yacht Club Commodore: "Gentlemen, congratulations. You have been selected to defend the America's Cup against "Australia" in the twenty-third Challenge Match." (Amidst throngs, cheers and strobe light flashes of cameras)

MR. AND MRS. EDWARD TURNER III DINE AT WHITE HOUSE

President Jimmy Carter: "I've been a very close reader of the sports page for the last several weeks, because we have a very distinguished Georgian who has, I think, come forward with a great deal of enthusiasm and skill, a great deal of understanding of the elements, the oceans in particular. He's exemplified, I think, the name of his boat. He's a very courageous man – Ted Turner. We are very proud to have you here tonight. And as you all know, he will represent us in the America's Cup races very shortly, having overwhelmed his opponents – much better than has been the case with his baseball team, the Atlanta Braves."

"COURAGEOUS" WINS FIRST RACE AGAINST ALL FOREIGN CONTENDERS

"I'm happy to be alive and to be from the United States and be able and healthy and fortunate enough to be participating in this great competition with my good friends on the 'Australia' and it's a little overwhelming to see all the nice people who have been so kind and everything..."

EYEWITNESS

Roger Vaughan (before second race between "Courageous" and "Australia"): "At least a dozen photographers had begged or stolen their way in, and were barely giving Turner walking room on the narrow floating docks alongside 'Courageous.' Turner didn't really have any specific dockside duties prior to casting off, but he was acting busy, shouting an order or two, studying things aloft, leaning on this or sitting on that, strolling, swaggering, strutting and posturing – all in the interest of good pictures.

"And the snapping of shutters and the whir of motordrives were all but drowning out the cries of seagulls on the neighboring fish docks. In his white shorts, white sweater with "Courageous" embroidered across the heart, his blue Greek fisherman's cap and dark glasses, he was this morning the living picture of the nickname 'Captain Courageous.'"

"COURAGEOUS" WINS SECOND RACE IN DUEL WITH "AUSTRALIA"

"Sometimes I think my father is somewhere watching all this. Watching me make the big time. I wish he could come back and see, you know? Like the father in 'Carousel?' The dead father, when he comes back to see his daughter at graduation?

"Damn! We were good friends. I wish my father could come back for just a day."

"COURAGEOUS" WINS THIRD RACE OUT OF FOUR

"All I know is I get what I want. Maybe because I want things more than others do. I wanted my TV station on the satellite, and there it is. I wanted to win the America's Cup, and that's getting close. I wanted to be worth fifty million, and in a few years it might be double that. People love me, all over the place, they really do. I can communicate on all levels."

EYEWITNESS

Bob Bavier: "The press loves Ted because he is such good copy, quick with one-liners, and never dull...

"In the long history of the America's Cup no skipper has done more to draw attention to the contest. Now not only yachtsmen – but cab drivers, ball players, rural housewifes and, in fact, people of all ages and walks of life know about the races. They may call it the 'American Cup' races, that sailboat race up at Newport, or some outlandish name, but they know about it as never before. And they care! Many misguided yachtsmen wanted the challenger to win, but the great American public wanted Ted Turner to beat the Australians. He was the people's choice."

ON ENORMOUS PUBLIC INTEREST IN "CAPTAIN COURAGEOUS"

"The reason I got really famous was that Bowie Kuhn suspended me from baseball that year. I'm not kidding. It's true. When I went off for the Cup trials it was as an innocent man serving time in the Commissioner of Baseball's jail. There's nothing like injustice to make people pay attention."

WHEN TOLD WHOEVER LOSES THE CUP FOR AMERICA WILL HAVE HIS HEAD PUT UNDER GLASS AT THE NEW YORK YACHT CLUB

"I sure as Hell don't want that to happen. But even if I lose, I don't think they'll take my head. They don't allow mustaches yet down at the New York Yacht Club."

EYEWITNESS AFTER "COURAGEOUS" WON THE FINAL RACE IN THE 23RD CHALLENGE MATCH OF THE AMERICA'S CUP

Roger Vaughan: "To be amid the scene was to be stunned. It was an armada gone mad, water-borne jubilation approaching hysteria. If 'Courageous' had made straight for Benton Reef, the ecstatic fleet might have followed like lemmings. Perhaps 400 vessels cavorted at top speed, rolling and surging and crashing along, bows slapping spray 40 feet in either direction, the throb of powerful engines a discordant earth rumble, the cheers and blowing of horns continuous.

"Overhead, small planes seemed to relive wartime dogfights as they vied for camera angles. The blimp descended like a dark cloud, her dangling mooring lines curving sensuously toward 'Courageous'' swaying mast tip. Coast Guard fire boats pumped towering plumes of water, filling the air with mist. Rising and falling on the swells as she plunged through the disturbed water as fast as she could go, 'Courageous' was a wanton, naked beauty in a lascivious dance before a whole flotilla of lust-crazed Turks...

"Every dock and pier along the waterfront was jammed with people. Every roof top had been turned into bleachers. Everything that could float had been cast into the harbor and manned. Cannon were fired. Horns blew in a cacophony of screeching...

"A colorful hot air balloon rose and drifted past on the breeze...

"Laughing and crying, people embraced." (Even the bells of Newport's churches were pealing their welcome.)

PANDEMONIUM ENSUES, AQUAVIT FLOWS

"Somebody put a bottle of aquavit in my hand and that's the last thing I remember..."

CAPTAIN COURAGEOUS – UH – OUTRAGEOUS

THREE SHEETS TO THE WIND

Rubbing bald head of press conference moderator while crew of "Courageous" sang "Dixie": "I never loved sailing against good friends any more than the French – uh – or the Aussies. I love them. They are the best of the best. The best of – the best... (Crew passes him a note) And I want to thank George Hinman of the New York Yacht Club for the opportunity.

"We worked hard. We busted our behinds. You are – all – the best of the best. (Now on floor under conference table. Reappears) And – the best – crew." (At this cue, the crew hoists their skipper high on their shoulders and carry him out into the night.)

LAST WORDS, ON BEING PUT TO BED

"Wouldn't the old man be proud of me tonight?"

EYEWITNESS

John Bertrand (future Australian America's Cup winner): "The New York Yacht Club should have installed a brass commemorative plaque on the spot where Turner fell, an appropriate memorial for this last great gesture of carefree amateur recklessness."

ON COMPETITION

(Later) "Sure I was drunk as a skunk on national TV, and you would be, too – right? I mean, we won! And we went through so much bull along the way, you wouldn't believe it. But we'd been beaten in 1974, and when we came back we wanted it bad the way underdogs always do. We could've quit after that, like Mohammed Ali did several times, but did we? Hell, no. We went right back there the next time, in 1980, and we gave Dennis Conner his shot against us. Competition is what it's all about. You beat the Hell out of somebody, and then you say, O.K., here's your chance to beat the Hell out of me. And then you beat the Hell out of him again – if you can."

ON WINNER'S SPEECH

"You ought to catch those Super Bowl winners when they have had several HOURS to enjoy victory instead of just a few minutes. Honestly, what was I going to say? A friend of mine was disappointed in me. He thought I'd missed my moment. My MOMENT? How the hell could I be profound? It was just a boat race. It was over. I had been away all summer. It was time to get back to work. I didn't even have time to make the Today Show, and they wanted me BAD. I have to work to earn a living.

"Anyway, my father always said not to set goals you can reach in your lifetime. After you accomplished them, there would be nothing left. I'll have other moments. I'll have more fun. My teams may keep losing but I'll enjoy myself."

HERO'S WELCOME AT ATLANTA AIRPORT

(Brass band plays Rocky) "I'm overwhelmed by all this... I wish I'd written a speech 'cause you all deserve some good words...

(Asked about some of his controversial behavior): "I don't mean to be a loudmouth, but I guess where there's smoke there's fire. I try to be good but sometimes it's hard..."

HEADLINES, "CAPTAIN OUTRAGEOUS IS HOME"

"After all, winning the America's Cup isn't as important as winning the World Series."

REFLECTING

"I had just gone on the air with the Superstation when I left for Newport, I had just bought the Braves, got suspended, and bought the Hawks. It was a marvel, really."

EYEWITNESS

Darrel Chaney, Braves shortstop: "I think he broke precedent one time by taking us all to dinner at The New York Yacht Club. So we all go walking down the street – here come the Atlanta Braves – we didn't take a cab or nothing – we walked. The whole team, Ted leading the way!

"The attire was 'coat-and-tie.' Nobody on our club even WORE a coat and tie very often... and Ted finally said, 'Oh, heck. Forget the ties. We'll make out alright. Let's go.'

"Once we were in it was like recess at school, you know, everybody walking around. Ted showed us all the boats and such – he even had the bar set up for us and everything...

"Even the waiters were first class – towel over the arm and all... the royal treatment – prime rib with horseradish sauce...

(Uses coathanger with rubber band gadget to sound like farts) "So – I let the thing go and said, 'Oh, my God, I'm truly sorry.' Turner got all excited. 'Who's doing that?' he yelled from the back of the room. 'You guys are SICK'...(same sound comes from Ted's own table).... 'Oh, no, Eddie (Robinson), not you, too. Everybody on the whole team is sick tonight!'"

1980 SPRING BANQUET, NEW YORK YACHT CLUB

"No matter who wins this summer, I will retain the America's Cup..." (Ted then unveiled the authenticated twin sister casting of the America's Cup trophy he had purchased from a British antique dealer.)

"COURAGEOUS" AND 1977 CREW LOSE 1980 CUP TRIALS TO DENNIS CONNER

"It's great to win and it's not so much fun to lose but it's not that big a deal. Christ, it's just a sailboat race." (Turner had just started CNN.)

EYEWITNESS

Dennis Conner, Ted's tactician in 1974 trials: "Ted's strong point is neither innate ability nor attention to detail and preparation – it's his enthusiastic competitiveness and leadership ability. He drives himself and his crew as hard as men can be pushed. This combatitiveness can be good some of the time and bad some of the time. Ted has a tendency to think only of the battle and not of the war. This may mean grinding down one opponent on the corner of the course while he forgets about the rest of the fleet. It may also mean steering for a dozen hours straight until he collapses from exhaustion. His kind of aggressive leadership works well when times are tough, but it can be counter-productive when things are going well."

EYEWITNESS

Bob Bavier: "Ted Turner's contribution to Cup awareness has to be rated at the very top. No one else has gotten the public so involved – no one else was such a sentimental favorite.

"True, he was drunk at the final press conference after 'Courageous' won in 1977, but how many baseball players are sober three hours after winning the World Series?

"When Ted Turner and 'Courageous' were the first to be eliminated in the 1980 defender trials, all of Newport was in virtual mourning. The people's choice was gone, never to return. Ted says, quite rightly, that three times is enough. But he will be missed, because in his inimitable way he has made more people be aware and care about

CAPTAIN COURAGEOUS – UH – OUTRAGEOUS

what happens in races for the America's Cup than anyone else in the event's long history.''

1980, THE FASTNET RACE
WHILE CNN IS GETTING ORGANIZED, TED GOES RACING OFF COWES, ENGLAND. DUE TO HUGE STORM, RACE BECOMES "FASTNET OF DEATH"

(Turner is reported missing on TV news for two hours, then appears on screen talking away):

'' 'Tenacious' was alright. It's those dishonest little things, skinned out hulls to save weight, that can't take it. There's got to be some legislation against them. And there has to be something done about the people who go to sea and haven't had the experience. Designers have to change the emphasis from speed to safety. This tragedy will bring about changes. But this situation should never have been allowed to exist.''

EYEWITNESS

Bunky Helfrich, while Ted was reported missing: ''I hope they strapped him below... At times like this, we used to call him 'Captain Panic,' and we would run the ship and strap him down below.''

"TENACIOUS" WAS "MISSING" BECAUSE THEY WERE SO FAR AHEAD

''My worst moment was when I was told some little boat was the winner. I had four hours of bitter disappointment before it was straightened out.

''It was a big sea alright, but we pressed on and never thought about stopping racing. One or two were seasick, but at the height of the storm we had a steak dinner.''

"It's no use crying. The King is dead. Long live the King. It had to happen sooner or later. We won because we had a good crew and a strong boat and a lot of experience, and the people who didn't have those went to the big regatta in the sky. I'm not going to say I'm sorry I won. I'm not going to say it."

"Like any experience, when you come through it you feel better. We're not talking about the other people who died, but to be able to face it all and come through it is exhilarating. Sailing in rough weather is what this sport is all about."

"Was I afraid? I guess I'm more afraid of being afraid than being afraid. But I was concerned. I wasn't too concerned about our survival, because they said the worst it would be is Force 10, and the waves are the biggest problem. You never know exactly what shape they're going to take. We got hit by quite a few that knocked our boat literally flat.

"You always feel bad when your fellow yachtsmen drown. But you never can be completely prepared for what nature has in store. We knew it was coming. We listened to the weather forecasts. But four people died on land and how can you prepare for something like that – trees falling and walls falling?"

BACK AT CNN

"It was rough, R-U-F-F. It wasn't a pleasure cruise but we had a good crew and a good, big boat. From what I can tell, it seems the smaller boats were the ones that got in trouble there."

"I remember saying to the crew that twenty men would die that night. Regretfully those turned out to be prophetic words.

CAPTAIN COURAGEOUS – UH – OUTRAGEOUS

"Despite being knocked flat six or seven times that night, my worst fear was that of running down smaller boats in the darkness. If we hit one of those lightly constructed fiberglass boats, 'Tenacious' would crush it, smash it in two, and everybody aboard would be killed. Of all the things that happened that night, that was my greatest fear, and that is the only thing that had me scared – that and the fear that something aboard 'Tenacious' might break, causing us to lose the race. You're supposed to have a strong vessel with crew and equipment for any condition. I feel a little like Noah. I knew that the flood was coming and I had a boat that would get me through it.

"It was a storm precisely like this one that saved England from the Spanish Armada. Whenever you sail in the English Channel, you've got to be prepared for the return of that storm."

AFTER GIVING UP OCEAN RACING

"I never did enjoy sailing that much."

CHAPTER SEVEN
TED TURNER DISCOVERS THE WORLD IS ROUND

TED FIRST HEARS ABOUT SATELLITE TRANSPONDERS FROM REESE SCHONFELD AND SID PIKE

TURNER: "What's it going to cost?"

REESE SCHONFELD: "It'll cost you a million dollars."

TURNER: "A million dollars to reach everybody in the country?"

SCHONFELD: "That's right, but you've got to understand they've got to put in their own dishes to receive your signal."

TURNER: "Oh, I understand that. I understand that. But do you mean that for one million dollars I'm going to be able to put this out all over the country?"

SCHONFELD: "Yeah, absolutely. Maybe a million one. Ninety-thousand a month. Something like that."

TURNER (later, to Sid Pike): "Sid, I want to buy an uplink. Send a salesman over and I'll get you a check."

IF ONE SATELLITE COULD BLANKET A CONTINENT, WHAT COULD FOUR DO – OR FIVE?

(Later) "If you were managing just one thing in the world, wouldn't the global flow of television news be about as big a thing – I mean the first global broadcast network?"

'ROUND THE WORLD, 'ROUND THE CLOCK

"I came up with the concept for the Cable News Network even before the Superstation was up on the satellite, because business is like a chess game and you have to look several moves ahead. Most people don't. They only think one move at a time. But any good chess player knows that when you're playing against a one-move player, you'll beat him every time.

"That's basically the way the networks (ABC, NBC, CBS) were moving. But I've always thought several moves ahead. I pride myself in being able to look into the future and say, 'What's the future going to look like? What can we do to be at the right spot at the right time?' Once it's obvious to everybody that something is going to be successful, then the opportunity is gone. Then anybody else can do it, too.

(Little ol' Channel 17 was about to send it's signal up to a "bird" and down across a continent, and be renamed "The Superstation") "It was clear that after the Superstation the next important service to the cable industry would be a twenty-four hour news channel. At that point, however, cable penetration was only at 14% of the TV homes in the country, so it wasn't economically feasible. There weren't enough cable subscribers to underwrite it."

"One of the greatest things I ever wrote was:

> 'INDECISION'
> 'At the feet of Hannibal
> Like a ripe plum Rome once lay
> Off he put the time of conquest
> To a later, better day.'

"Pretty strong, huh?"

FCC THREATENS TO RESTRICT SATELLITE TRANSMISSION. TURNER IS ASKED TO TESTIFY ON BEHALF OF THE CABLE INDUSTRY

At Senate Subcommittee on Communications hearing, July 1976: "My name is R.E. Turner III. I am chairman of the board of Turner Communications Corporation, licensee of UHF television stations; WTCG, Channel 17 in Atlanta, Georgia, and WRET, Channel 36, in Charlotte, North Carolina. I am also president of the Atlanta Braves National League baseball club. Turner Communications also has holdings in a marketing company, an outdoor advertising firm and several radio stations.

"I purchased Channel 17 about six years ago. At that time the station had only a small off-air viewing audience and was received on only a handful of cable television systems.

"A UHF independent television station in a market such as Atlanta where there are three VHF network-affiliated stations, has to do a lot of scratching to attract audiences that are going to attract advertisers to buy commercial time.

"I put a lot of money into improving the technical quality of WTCG. I put a lot of money into buying the best programming available. And I secured the television rights to Atlanta's three major league professional sports teams in basketball, baseball and hockey.

"And I did one thing more. I visited the cable systems. I attended their association meetings, and I told them about all the things we were doing with Channel 17. Today, nearly 100 cable television systems in five states bring WTCG to more than 400,000 cable TV subscribers. A lot of the success we are beginning to enjoy at Channel 17 is due to the cable television audiences.

"I am unique, I think, or fairly unique, in being friendly to the people in the cable industry and I'm considered a bit of a traitor to be here on behalf of the cable industry this morning, 'though I do feel I have a pretty reasonable position.

"Once you get outside Atlanta, there are very, very few independent television stations in the South. The only commercial television available to most people off the air is from the three national networks out of New York City. The people in Columbus, Georgia or Knoxville, Tennessee or Tallahassee, Florida and in all the little towns in between – the only programming choice they have is what the national networks choose for them to see.

"I came into the independent television business because I believe there should be more voices to be heard than the network voices out of New York and more opportunity for program selection by the American people. In most towns and cities in the South, there are no allocations available for local television stations. And, even if there were, most of the towns couldn't support another commercial television station.

"So the only chance for most people in the South to get a choice of programming is to let cable operators bring in an increased number of distant independent signals like WTCG from Atlanta.

"My hat is off to the cable people, and I'm pulling for them, I really am. They are providing a real public service for the American people. There should not be any distant signal carrier restrictions for cable TV. To limit the number of television stations available to cable systems is to short-change the American people and to perpetuate a broadcast monopoly.

"Especially, this year of all years (the Bicentennial), we should be promoting and fostering the widest possible freedom of choice.

(Is your station known as Super 17?) "Well, that was more or less a joke. I got that name six years ago when we were losing eighty thousand dollars a month and were watched by no one. We had a young girl down in promotion who did not last very long, but she had one great idea. I said, 'We need to jazz this place up a little bit.' And she said, 'Why don't we call this place Super 17?' I said, 'That's a great idea!' You know, in other words, it was a real tongue-in-cheek thing, and it was long before the Superstation concept...

"Well, I would love to become a Superstation. I would love desperately to create a fourth network for cable television, producing our own programs, not just running 'I Love Lucy' and 'Gilligan's Island' for the fifty-seventh time. And I intend to go that way if we are allowed to. You have to remember, there are three supernetworks, who only own four or five stations each, that are controlling the way this nation thinks, and raking off exorbitant profits, and most of these local stations that everybody is crying about are just carrying those network programs that originate out of New York. They have an absolute, a virtual stranglehold on what Americans see and think, and I think a lot of times they do not operate in the public good, showing overemphasis on murders and violence and so forth.

"So if we do become Super it will be another voice. Perhaps it will be a little more representative of what we think the average American would like to see. A little less blood and gore on television and a little more sports and old movies and that sort of thing, that we might encourage children not to go out and buy a gun and start blasting people like in 'Taxi Driver.'"

AFTER CONGRESSIONAL HEARING

"After I debated Gene Jankowski (CBS Broadcast Group President) he crawled out on his hands and knees."

TO HIS WASHINGTON LOBBYIST

"I want to live five lives. I have to hurry to get them all in."

THE SUPERSTATION RECEIVES FCC LICENSE AND SENDS ITS SIGNAL ACROSS A CONTINENT THE SAME DAY

"It's going to be tough, but we'll do it. Someday you four guys (Jim Trahey, Jerry Hogan, Bob Sieber, Don Lachowski) are going to make a whole lot of money."

TO STAFF

"Problems are just opportunities waiting to be seized."

THEY CALLED THEMSELVES "TED TURNER AND HIS MERRY MEN"

"Look what's happened in a year! In terms of cable homes, we've done in one year on the satellite what it took seven years to do on the microwave."

THE WAR BETWEEN THE NETWORKS AND SATELLITE CABLE BEGINS

(Brandishing a confederate saber and pacing in quick circles) "The networks are a bunch of pinkos! (Slicing the air) In their race for ratings, their newscasts dig up the most sordid things that human beings do, or the biggest disasters, and try to make them as exciting as possible.

TED TURNER DISCOVERS THE WORLD IS ROUND

"In their entertainment programs, they make heros of criminals and glamorize violence. They've polluted our minds and our children's' minds. I think they're almost guilty of manslaughter."

"The FCC had to change the rules a little bit and now we've got everybody in the world suing us. The networks are scared to death of cable television and now that we're on satellite, they're REALLY scared. We're sucking up the market.

"But NBC, ABC, CBS, the Motion Picture Association of America, the NBA, the Baseball Commissioner's office, the BBC, the National Hockey League and an assortment of other people are trying to stop us.

"Can you imagine that? A little old raggedy station with a hundred employees and a bunch of torn-up furniture is going to destroy television and cause the motion picture industry to collapse."

LIKE ALEXANDER THE GREAT, TED HAS NEW LANDS TO CONQUER

"I can do more today in communications than any conqueror ever could have done. I want to be the hero of my country...

"I want to turn it around before it is too late. And the hour IS late. Someday, somebody will put a bullet in me. I would like to stay around for a while, but I really do believe I'll be assassinated."

SUCCESS?

"Don't be a reckless gambler. Take risks but only after preparing. And don't put all your eggs in one basket."

ON COVER OF SUCCESS MAGAZINE

(Waving magazine over heads during speech at Georgetown University) "Is this enough? Is this enough for you, Dad?"

"My father died when I was 24. That left me alone, because I had counted on him to make the judgement of whether or not I was a success.

THE FUTURE?

"I have four great ambitions. One, I'm going to make Channel 17 the fourth national network. Two, I'm going into the production business – they're producing trash on movies and TV. Three, I'm going to be the country's wealthiest man. And four, I'm going to be the President of the United States.

(With no political base?) "I've got the boob tube. If this country falls flat on its face, I can go on the boob tube. That's power."

(Later) "I would only run for president if it was the only way I could get this country to turn around. My main concern is to be a benefit to the world, to build up a global communications system that helps humanity come together, to control population, to stop the arms race, to preserve our environment. I'm a deep thinker. I've traveled all over. I have more access to information than anyone on the planet. When you realize your family, your friends, your society, and your planet is in a dire state of emergency, that has to change anyone with a responsible world outlook.

"I've thought about being president from time to time, and people have asked me about it from time to time, but I like my present job a lot more. I said back in the early 1980's that I wanted to be Jiminy Cricket for America. You know, the country's conscience.

"Some day there will be global networking, and we're set to be the leaders in that."

TED TURNER DISCOVERS THE WORLD IS ROUND

WITHIN A FEW YEARS ATLANTA'S TBS WAS TRULY 'ROUND THE WORLD, 'ROUND THE CLOCK

"We're all over the world. We are just as capable of coming up with something out of Moscow or Frankfurt or London as we are here in Atlanta. But you've got to have somewhere that you're headquartered, and our headquarters is here.

"There are a lot of advantages to being here, as opposed to New York. We're able to operate a lot less expensively than the New York based networks... I had people from the big networks say there was no way you could do a network from Atlanta. They said it's unthinkable. So I said, 'The world's largest soft drink company operates out of Atlanta.'

"I said, 'You can do it anywhere.' We've certainly exploded that myth."

(Later adds two news networks) "I met with the head of Indian television when I was there – there's only one network in India, 750 million people - and they said, 'absolutely' they'll participate (in CNN's World News Report). The Third World is very upset that all the news agencies are controlled by the West, and that they have no input. The news that goes around the world on television is pre-edited. The only time the Third World ever gets any coverage is when there's a Bhopal or riots or something like that, and they're upset about it.

"Why should the BBC and CBS decide what the news is in the world? Two thirds of the people live in the Third World. We may change the whole way things are done around this world. We may find out that it's pretty nice to get the news and ask people questions directly, as we did to the Iranian ambassador to the UN. It's better than having Casper Weinberger telling us what Iran is all about and what the problems are there. Why not get the news from the horse's mouth? These things are coming to me as I study it, and they're being well received."

"There's still a lot of room to grow in Europe and the rest of the world. And everybody is going to be able to use 'World News Report,' which we're going to start this fall. That's really going to move us up a notch as far as the world is concerned. The 'World News Report' will in time probably be expanded into an agency where the news is fed in directly by the countries and then goes out. Why shouldn't it have been like that in the first place?"

(After buying MGM movies) "The movies that we have, 'The Wizard of Oz' and 'Gone with the Wind' and 'Ben Hur' – people can understand those movies everywhere. We own (the cartoon) 'Tom and Jerry' and there aren't even any words in it. Ten years ago the Soviet television was running 'Tom and Jerry.' But you can run 'Tom and Jerry' and you don't even have to dub it into languages because all it is is music."

"We have the biggest and best library of film that exists anywhere – films that the whole world would like to see. Whereas the whole world wouldn't want to see 'Miami Vice' or 'Knots Landing' because sitcoms don't play worldwide. Movies play worldwide."

CONQUERING THE NEW WORLD HE DISCOVERED

"I'm very ambitious, and if I wasn't we wouldn't be where we are now, our hope is to be the leading producer and distributor of news and information and entertainment programming to television audiences around the world. One thing that people don't understand is that money has never been what motivates me. I was willing to take chances because I didn't think the money was that important.

"The challenge and the adventure were the main things with me, and the sense of achievement and accomplishment, and wanting to do something in my life that would be really spectacular. I've always had grandiose schemes."

THE POWER TO PERSUADE AND INFORM ON A GLOBAL SCALE

"People have got to be better informed of where we're heading. It would take only a billion dollars a year to furnish birth control devices to all the women in the world who would use them. That would cut the world population growth in half. One billion – that's about the cost of one Trident submarine, one three-hundredth of our current military budget.

"We're steaming at 30 knots on the Titanic trying to break the trans-Atlantic record on an iceberg strewn sea. We're out of control!

"WE'VE GOT TO GET IN CONTROL!"

CHAPTER EIGHT
CNN CREATES A GLOBAL TALKATHON

BUYS 5,000 ACRE HOPE PLANTATION

"It's the only thing that lasts, land. Land and the video business will last forever."

WALKING IN THE EERIE FOREST UNDER THE SOUGHING PINES

"We walk with the ghosts of five hundred slaves. They worked the rice fields well into the nineteenth century – there were three plantation houses then, not just one. Hope Plantation seems like a sleepy place, but things aren't always what they seem.

"All around us you see the survival of the fittest going on. Every little animal competing for what's his. It's natural. You can learn a lot from it...

"We don't have to get fat and lazy. Let's have fun again, let's stick up for ourselves. Look at me. I'm about to be the fourth network. We already are, in fact, with Channel 17 in Atlanta. I coined the word Superstation. We got on the satellite first, we're in nine million American homes right now where only the networks were before.

"And do you know why we're going to be such a big success and why I'm going to make a billion dollars? It's because people are all screwed up, and they're looking for a change. It's not that I'm a genius. It's that the three networks have it all to themselves and they want to keep it lousy.

"The networks are like the Mafia. The networks ARE the Mafia. Do you know they spent a quarter of a million dollars in Washington trying to stop my Superstation from showing movies and sports in people's houses? Well – their day is finished now. It's over. They've made unbelievable profits, and what have they brought us? Mr. Whipple squeezing the toilet paper. The $1.98 Cheap Show. The Newlywed Game. Love Boat.

"I'm telling you the networks are scared. But they can't stop me because people are demanding – they're insisting – on alternatives.

"There's an indigo bunting. I've got two baby bison, and one of them was born last week... and baby geese, baby turkeys, plus I have two bears and there's alligators, and we're raising all kinds of ducks. I'm making a Garden of Eden. It's amazing how tame things will get when you're not trying to kill them.

"Countries should act the same."

"In 1978 I started thinking about an all news TV channel again."

"News is really in the Dark Ages compared to movies, sports, and general programming."

MARKET SURVEY FOR CNN – TO CREW OF "COURAGEOUS"

"What would you guys think if you could turn on your TV any time of the day or night and find out what's happening in the world?"

MARKET SURVEY FOR CNN - TO CAB DRIVER

"Don't you think an all news TV channel would be terrific?"

CNN CREATES A GLOBAL TALKATHON

TURNER SEES SATELLITE NEWS AS THE NEW WORLD, A TERRITORY TO BE CLAIMED

"By then several other services had gotten started, but they were very weakly funded. The cable industry was growing mainly with HBO and the Superstation and Showtime, which was now in existence. But still no one had tackled news.

"Even though the Superstation had not really become a success with the advertisers yet, I could see that it was going to become a success. In my opinion, we were going to become successful selling cable advertising – and without that you couldn't have afforded the vast programming expenses that were going to be necessary to do an all news channel. That's why no one else had stepped forward. The easiest way would have been for one of the three networks to do it, but they didn't want to undermine their affiliate system. And it still wasn't clear that cable was going to be a success. They didn't WANT it to become successful. They loved having just a three channel environment."

AT COCKTAIL PARTY IN FRONT OF A CROWD, PLAYIN' REAL DUMB AND ACTIN' LIKE A FOX

REESE SCHONFELD: "Hi, Ted. I was wondering if you wanted to buy some news for your station."

TURNER (in loud voice): "NEWS? Ah, Reese. Ah can't do news! Ah got too much other product right now. And who wants news, anyway? News is nothin'! NOBODY watches news. Ain't you TIRED of all that news? Don't it just make you SICK after watchin' all that stuff? Listen, Reese. You know what my motto is? No news is good news."

TURNER MAKES LIGHTNING ATTACK

"Lachowsky! Get a yellow pad and come in here! I'm going to do a twenty-four hour news network. Write this down. I'll tell you what it's going to be. It's going to be a half hour of news like Time

Magazine. Then a half hour of sports like Sports Illustrated. Another half hour of features like People Magazine. And a half hour of business news like Fortune. We're going to repeat it every two hours, twenty-four hours a day. We're going to do this half hour format and freshen it up every two hours."

SHORTEST AND MOST IMPORTANT CALL TURNER EVER MADE

(To Schonfeld) "Hello, Reese? I'm going to start a twenty-four hour news channel. Can it be done? (Yes) Will you do it? (Yes) O.K., then. Come on down to Atlanta as soon as you can so we can work it out."

SCHONFELD ARRIVES IN TED'S TROPHY-FILLED OFFICE

TURNER: "There are only four things that television does, Reese. It does movies, and HBO has beaten me to that. It does sports, and now ESPN's got that. There's the regular series kind of stuff and the networks have beaten me to that. All that's left is news! And I've got to get there before somebody else does, or I'm going to be shut out. I'm going to call it the Cable News Network, not the Turner News Network – the Cable News Network – to encourage the support of the cable industry. This is the one thing I've got to bring them all into the tent, so I'm going to call it CNN." (Shows him format on yellow pad)

SCHONFELD: "Well, Ted, I think the format has to be more flexible, to accommodate more breaking news, more live coverage, more up-to-the-minute news. Our thing is going to be our ability to get to the news first, put it on the air while it's happening. We'll be the only ones in the world who can do that."

TURNER: "Right!"

TRYING TO HIRE DAN RATHER

"If something's going on anywhere in the world, we're going to bring it right in live. People want to see it happening right in front of their eyes."

SCOUTS LOCATION FOR CNN BUILDING WITH BUNKY HELFRICH, ARCHITECT AND CREWMATE ON "COURAGEOUS"

"Come on with me. I want to show you some things I've got on my mind."

AFTER CHOOSING A FORMER JEWISH COUNTRY CLUB WITH WHITE "GONE WITH THE WIND" STYLE COLUMNS

Bunky Helfrich: "I knew there was not enough time to do the job (extensive renovations), but there was no use telling Ted that. You just did it."

EYEWITNESS

Will Sanders: (Ted's gone off sailing again in Fastnet Race) "He stirs up so much turmoil that if he doesn't stay away for extended periods of time the whole place might collapse. During the Cup summer, for instance, he spent only ten days here. That was fine. I told him once when he was worrying about money that if he would agree to take a one-year world cruise, he would be a rich man when he returned.

"He is swinging away all the time. He does business wherever he meets people. He'll meet a guy on an airplane. Then he'll call: how come we don't do this or that?

"Once in a while the dust has to settle. Occasionally you need smooth water to tidy up the ship."

WORD OF STARTING CNN GETS AROUND SUPERSTATION

(Anonymous sign appears on Ted's desk)

> "PLEASE, TED!
> DON'T DO THIS TO US!
> IF YOU COMMIT TO A VENTURE OF THIS SIZE
> YOU'LL SINK THE WHOLE COMPANY"

OPERATING OUT OF OLD WHITE (HAUNTED) HOUSE WHILE FUTURE HEADQUARTERS WERE UNDER RENOVATION:

Eyewitness, Dee Woods (who screened all Ted's visitors): "I had met Ted back in April 1976 before we went on satellite with the Superstation, and I'd learned since then that once he makes a decision he never looks back. He just continues to look forward, not necessarily just to next year but to five or ten years from now – and what he does, what he's really good about doing is searching out the best people to do a job – so that then, instead of standing around, he can just let them do it. I think that is what he was doing in gathering these individuals over there in the haunted house. He had gotten the best people available to do Cable News Network and now they were getting everything organized and pulling it together."

EYEWITNESS

Mary Alice Williams: "It was hilarious. There was a card table in the room where we worked and we'd have coffee cups setting there, and every time a bus would go by the table would shake and the cups would go flying all over.

"There was so much laughing. But our loyalty and our spirit and our guts went into it. We sat there trying to figure out how we were going to fill twenty-four hours of television news with credible and reliable information, every single day, without any role model, and with everybody on the planet having asserted for the last two decades that it couldn't be done."

(Her new boss, Ted Turner?) "His mood swings were so wild. I mean, he was so loud, he was so high – every normal human emotion was magnified with Ted. He was just so much larger than life. And I suppose at first I thought he was just as crazy as a bed bug."

BUT IT WAS FUN WORKING FOR TED

Eyewitness, Gerry Hogan: "He was selling US. His selling ability in advertising is more than adequate, but he's really a great promoter and a great motivater. It's his personality. It's the kind of leadership the military uses to get young troops to walk into a wall of fire. He had that ability to inspire you and commit you to a higher cause."

CNN'S ORIGINAL STAFF OF 300 WAS LEAN AND MEAN

Ira Miskin: "We were all wounded soldiers, those people who came to Turner from the mid-1970's to the mid 1980's...

"All of us who worked for Ted at all the various levels were no saints. We were nuts. We were the kinds of kids who should be put in the corner for being pains in the asses – cage rattlers...

"We were all strident, lean, mean, foul, pushy. It was all part of the internal persona, and it kept the pot boiling. It made the company work – because there were so many crazy, psycho, committed people."

UNDERGROUND GUERRILLA WARFARE AGAINST THE NETWORKS

Eyewitness, Jim Shephard: "We were cocky and arrogant and overbearing. We drank too hard and we worked too hard, and the place was full of all these relationships and affairs that were hotter than the surface of the sun for thirty days before they burned out. We were intense. One dimensional. Driven.

"And we were nuts for doing what we were doing. We used to work seventy or eighty hours a week and still go out and drink all night. We were bad for that. We unwound the only way we could.

"Because we had a crazy man named Schonfeld, who would come downstairs and go whacko at a moment's notice, and another crazy man in residence named Kavanau, and then all the rest of us who were selectively nuts anyway.

"I mean, we were the dissidents. We were the people who would never have succeeded at traditional places because we were too hard to control. We were headstrong and impetuous. But because of that, we were perfect for CNN.

"We were perfect for guerilla warfare, because you could never teach us to march in step. It was like taking the Dirty Dozen and putting them together until they became a unit."

AT CNN, NO STARS BUT THE NEWS?

Lois Hart: "Reese Schonfeld was the guy who really made CNN. He's a journalist at heart, even though he was trained as a lawyer and seemed to spend most of his time as an administrator. Reese is also crazy, wild, with a monumental temper. None of us had any idea how we were going to fill up twenty-four hours, but to Reese that was never a problem. He worried about what it was all going to look like on the screen – what kind of format CNN would ultimately have. This may sound ridiculous, but I don't think any of them – not Reese,

CNN CREATES A GLOBAL TALKATHON

not Turner, not Kavanau, or any of the others who were responsible for getting this incredible thing off the ground – I don't think any of them really knew what CNN would look like. The one mantra that everybody chanted during those early days was 'Nobody's a star. The news is the star.' Well, that was fine, but we all knew the real star was Ted Turner."

BLEAK DAYS

"Nobody thinks we can do it. I tell them, 'Tough, we've got to.' And they don't even know the problems. I based our entire startup funds on the sale of the Charlotte station to Westinghouse – that's my twenty million dollar ante right there. But the sale hasn't even gone through yet, because there's a group down there that's contesting the license.

"I've got Bunky Helfrich working full-time redoing the new headquarters here, and that depends on a ten million dollar bond issue that the city of Atlanta hasn't even approved yet.

"We haven't figured out a way to bring satellite pictures to Atlanta, which is where all our news has to come before it can go back out. Terry McGuirk is trying to sell this new service to cable systems while all this is going on – without anything to show them except the promo tape. He's just blowing smoke, a huge amount of smoke.

"And we're hiring people right and left. Schonfeld already signed on about fifty, and we'll have a staff of three hundred before we click on. If any of this stuff goes bad on me, I'll be up the creek. It's going to be 'Good-bye Ted Turner, it was nice to know you.'"

SATCOM III DISAPPEARS AFTER NASA LAUNCH DASHING CNN'S HOPES FOR A TRANSPONDER

"I didn't know that satellites failed. I thought this thing was routine."

(Turner was in Nassau racing "Tenacious") "I'd just got off the boat. We'd raced over there from Miami, and we'd only been in for an hour. As soon as I heard, I grabbed Janie and we got the next plane home."

CBS EVENING NEWS

Walter Cronkite waxed poetic:

"T'was three weeks before Christmas
And down at the Cape
The Satcom III satellite
Seemed in great shape.

It was RCA's baby
That NASA would fling
As it happened
The one-hundred fiftieth thing

To be launched by a Delta,
A rocket so flyable
That's considered to be –
ABSOLUTELY RELIABLE

And from somewhere in space
Comes the seasonal call
Merry Christmas, Goodnight
And you can't win 'em all."

CNN STAFF

"We're dead!"

TURNER TO CNN STAFF

"Carry on."

WAS NBC TRYING TO KEEP CNN OFF THE AIR?

Background: "RCA had two transponders available on Sat com I, but six channels now wanted them. Ted and his troops were worried that since RCA owned NBC, they might try to block CNN from the satellite, and do their own news network. RCA had decided rather than satisfy two customers and offend four, they would lease no transponders at all."

AT WAR WITH RCA TO GET A TRANSPONDER

(Going up the elevator at Rockefeller Center with Reese Schonfeld and lawyers, Ted was a gathering storm cloud) Bursting into RCA offices:

"All you guys get outta here! I want the chairman of the overall parent corporation down here, right now, because I'm going to break this company into so many small pieces that all of you will be looking for jobs." (Turner was just warming up)

EYEWITNESS

Harold Ingles, RCA: "Turner ranted, he screamed, and he threatened. He accused us of taking this action to protect NBC from competition. He was going to go to the FCC, to the Justice Department, to Congress, and to the public. When he got done with us, he said, NBC would lose all of its licenses."

TURNER IS NOW WARMED UP

Grabs Ingles by neck of shirt, hoisting him up to nose level: "Andy, do you own any RCA stock? (Yeah, I own some) Well – you better sell it. When I get done with you, it ain't gonna be worth a dollar a share! We're gonna go after NBC..."

EYEWITNESS

Carl Cangelosi, RCA attorney: "I remember Ted yelling at me for about twenty minutes straight about how he was going to sue NBC for antitrust. How he was going to humble RCA Corporation. He was almost uncontrollable."

RCA CLAIMS CNN HAS NO LEGAL GROUND

"You're gonna kill me. The networks are gonna kill me. This is my death if you do this to me. This is my blood you're getting. I'm a small company and you guys may put me out of business, BUT – (And, after a long, deafening pause he roared):

"FOR EVERY DROP OF BLOOD I SHED
YOU WILL SHED A BARREL!"

EYEWITNESS

Reese Schonfeld: "Cangelosi made Ted this offer: 'We would assign him a transponder on Satcom 1 temporarily, while the court was reaching its decision as to our obligations under the 1976 contract. If, as far as we expected, the court ruled for us, he would be obliged to move his traffic to another satellite (without HBO, requiring stations to invest in a separate dish) on December 1, 1990. If the court ruled in his favor, he would keep the transponder. In return for this, Turner would agree to waive all claims for damages.'

"Ted was elated. He jumped up, ran over to Ingles, began pumping his hand up and down, talking all the while. He was saying things like there really was no conflict of interest... RCA should be happy... etc., etc. Tench Coxe said, 'Ted, shut up,' and they all got him in the elevator before he could do anything else crazy. The elevator was full of RCA employees and Ted's mouth was still running. Terry McGuirk and Ed Turner tried to talk even more to drown him out. He was saying things like, 'We made a secret deal...' He went on and on about how he had beaten down RCA."

BRANDISHING BROADSWORD OVER HEAD AT STAFF MEETING

Back in Atlanta: "We will not be stopped! No matter what it costs, we're going on!"

EYEWITNESS

Ted Kavanau: "Turner called us all into his office when he got back, and we figured it was all over. He is a terrific fighter, but he knew this time he was really backed into a corner. With everything on the line, he brought us over there to tell us he wasn't giving up. We all filed into his office half expecting some kind of funeral oration, and there he was, strutting around, pounding his fist, and telling us not to give up the fight. He made it seem almost like a crusade, only the enemy wasn't RCA. It was the networks. He was waving this huge broadsword he kept in his office, swinging it around over his head shouting, 'We will not be stopped! No matter what it costs, we're going on!' And after that there wasn't one of us in that room who wouldn't lie down and die for the guy."

ON RECEIVING A FEDERAL CONSENT ORDER FOR A SATCOM I TRANSPONDER FOR SIX MONTHS

Calls press conference amidst the rubble of the new building, giving everyone a yellow hard hat with CNN stamped on it: "This is not the solution we would have liked to have had. But it's the best we could have done. It will allow us to get off on time.

"Barring satellite problems, we won't be signing off until the world ends. We'll be on, and we will COVER the end of the world, live, and that will be our last event. We'll play the National Anthem only one time, on the first of June; and when the end of the world comes, we'll play 'Nearer My God to Thee' before we sign off."

TED HAD ALREADY TAPED THIS LAST EVENT

"Normally, when a television station begins and ends the broadcast day it signs on and off by playing the National Anthem, but with CNN – a twenty-four-hour-a-day channel - we would only sign off once, and I KNEW what that would mean. So we got the combined Armed Forces marching bands together – the Army, Navy, Marine and Air Force bands – and took them out to the old CNN headquarters, and we had them practice the National Anthem for a videotaping. Then, as things cranked up, I asked if they'd play 'Nearer, My God, to Thee' to put on videotape just in case the world ever came to an end. That would be the last thing CNN played before we – before we signed off. And, I'll tell you, those guys in the military bands knew what I was up to.

"I can't watch this (video) without getting tears in my eyes...

"Pretty strong, huh? I keep this tape around because when the world ends it'll be over before we can say what we wanted to say. Before we can leave any final messages."

HIRING CONTINUES

Ted Kavanau, running over budget: "I said, here's what we'll do. We won't hire these professional people. We will use the same budget, but instead of paying them $20,000 to $22,000 apiece, we'll go around the country to colleges. We'll hire the best kids coming out of these schools, pay them half the money and get twice the number. So we brought in this massive bunch of raw kids and tried to house them. We threw them into boarding houses and they lived together. These hoards of kids descending upon Atlanta... With less than two months to go, we started training an army of youth. What we had on our hands at the final hour, was a children's crusade."

EYEWITNESS

Pat Gorman: "We had then only a few seasoned professionals leading an army of neophytes and were faced with having to teach these kids how to USE the equipment just to get on the air."

TED AT LUNCH WITH LAWYERS

"This is about it. The whole deal is crumbling, and when it goes, it'll take everything I've got with it. I'm just about flat broke.

"We haven't got a (permanent) transponder and we haven't got the Charlotte money ($20 million from sale of the Charlotte TV station to Westinghouse).

"The banks are calling in their notes on me, and the insurance company already has. I've got three hundred people on the Cable News payroll, and no money coming in to pay them.

"I just had to borrow $20 million to tide me over. The interest rate is twenty-five percent. Twenty-five percent of $20 million is $5 million a year, and there's twelve months in a year and that's four hundred thousand a month in interest alone. I can't pay it."

"I'm going to collapse. The only question is will it be tomorrow, next week, or next year? I mean, I'm just so tired I can't make it."

TURNER GOES AFTER CHARLOTTE MONEY HELD UP BY BLACK COALITION CLAIMING UNFAIR HIRING PRACTICES

(Takes Hank Aaron with him) "You know, I don't blame you guys for being mad at me. I'd be mad at me, too...

"But it looks like you got the same problem I got in my company. You don't have any blacks in high places either.

"You got three guys here who're doing all the talking – and they're white!"

"They agreed to stop opposing the sale of the station, and we made some concessions. We made a lot of consessions. In fact, that little trip cost me close to half a million dollars." (But CNN is back in business with $20 million to fund its first year.)

CIRCULATING WITH REPORTERS ON NEWLY LAID SOD UNDER TENTS AT CNN LAUNCH, JUNE 1, 1980

"I'm going to do news like the world has never seen news before. This will be the most significant achievement in the annals of journalism."

"I'm going to travel around to every foreign country and get the head of the country to show me the things he's proudest of about his country. Then send it all back by satellite."

EYEWITNESS

Reese Schonfeld: "Nobody believed him, did they? With anybody else you would say that kind of talk is ridiculous. But Ted will find a way to do it. He was already discussing with me the possibility of getting a feed from Russian television in 1981. I told him I doubted their government would permit it, but who was going to tell Ted it was impossible?"

SPEECH AT LAUNCH OF CNN

"I'd like to call our ceremonies to order. We should be on the air at 6:00 o'clock as predicted. We have a few statements from some visitors and from some people who are with us...

CNN CREATES A GLOBAL TALKATHON

"You'll notice that out in front of me we've raised three flags. One is of the State of Georgia where we're located. In the center is the flag of the United States which represents the nation this Cable News Network intends to serve. On the other side is the flag of the United Nations because we hope, with our greater depth, and our international coverage, to make possible a better understanding of how people from different nations can live and work, and so to bring together in brotherhood and kindness and peace the people of this nation and world. I'm now going to read a little poem that was written in dedication:

"To act on one's convictions while others wait
To create a positive force in a world
   where cynics abound
To provide information to people
   where it wasn't available before
To offer those who want it a choice
For the American people,
   whose thirst for understanding
   and a better life made this venture possible
For the cable industry whose pioneering spirit caused
   this great step forward in communications
And for those employees of Turner Broadcasting
   whose total commitment to their company
   has brought us together today
I dedicate the news channel for America,
   the Cable News Network.

"Now if we can have the presentation of the colors and the National Anthem. Everybody please stand.

(Four military bands play the Star Spangled Banner.) "Awwriight! At least I think we're on the air. At least I hope so."

DASHING INSIDE NEW CNN HEADQUARTERS

"Hey, Reese. Is there any place I can watch the Braves game? They're on the Superstation now and they're beating the Dodgers 9-5. (Braves win 9-5) Awwriight!"

WHEN CNN WAS LAUNCHED TURNER MADE COVER OF NEWSWEEK

Cover story: "Much like Muhammad Ali, Ted Turner has enlivened and enriched all the games he has entered. Though he may occasionally mistake himself for one of his military heros, his sheer exuberance is always infectious; the relish makes up for the hot-dogging. More important, in an age of play-it-safe corporate beaurocracy, bold spirits like Turner have become precious commodities...

"Whether twenty-four hour news is an idea whose time has come is considerably less clear. Aside from the question of viewer receptivity, and the problem of cable's limited market penetration, there loom serious doubts about Turner's financial staying power. Some experts are convinced that CNN will eventually be forced to surrender the all-news cable field to journalistic organizations with deeper monetary resources...

"Maybe so, but consider what Turner is achieving right now. He is pushing television news to its farthest frontiers. He is providing viewers with an important new option. He may even galvanize the networks into expanding their own newscasts beyond what many critics dismiss as a superficial headline service. As for CNN's chances for survival, the fact that it is Ted Turner's current crusade may be reason enough to anticipate success.

"George Babick, head of CNN's New York sales office, offers perhaps the wisest advice about how to regard anything his boss touches: 'If Ted predicted the sun will come up in the west tomorrow

CNN CREATES A GLOBAL TALKATHON

morning, you'd laugh and say he's full of it. But you'd still set the alarm. You wouldn't want to miss the miracle.'

"Turner's latest 'miracle' rose last week. It will be worth watching."

AMIDST RAVE REVIEWS, HOME VIDEO MAGAZINE SUGGESTS CNN LACKS FOCUS

"Wait a minute! Wait just a goddamn minute! You think it lacks focus – what is focus anyway? If you're live all the time, how can you have focus? Focus means that you know where you're going! You can't focus on something unless you know what you're focusing on! Focus is something a newspaper has, because there's a day to think about it. Or with a magazine there's a month.

"Whoever said that is a yo-yo!"

CHAPTER NINE
BOY ARE THOSE FOREIGN KINGS AND PRESIDENTS GOING TO BE SURPRISED TO SEE ME!

BACK ON HOPE PLANTATION

"What I've got to do now is broaden myself and I've got a plan. I'm going to set up a receiving station here on the plantation to confer with the best and smartest people there are...

"But I'm also going to make my own kind of pilgrimage around the world. There's so much I don't know, but I'm going to find it out. I'm going to visit every country that will let me in, in Europe, and Asia and Africa, wherever they have a lot of problems. I'm going to meet with the leaders and I'm going to find out what they're thinking. Reading only gives you so much, but if you actually go there you can figure it out. Boy are those foreign kings and presidents going to be surprised to see me! It'll be interesting as Hell for both of us."

EYEWITNESS

Robert Wussler: "Ted Turner learned more between the ages of forty and fifty than anybody else I have ever met in my life. He asked a lot of questions, read a lot, looked a lot at TV. Here was a guy who didn't know everything, but was extremely eager to learn. He was a sponge, soaking everything up. I saw that man get more and more sophisticated. At a time when most people's learning curve goes down, his went up.

"He was learning a great deal about life, about style, about business. He became worldly, he traveled to China, Japan, the Soviet Union... (also Africa, Scotland, England, Denmark, Brazil, Mexico, Jordan, Israel, Cuba, Egypt, India, Australia, Tahiti, Bermuda, the Bahamas, Greece, Mikanos, Seriphos, and all across the US including Alaska and Hawaii.) While his company was growing by leaps and bounds, he was, too."

SETS HIS OWN ROLE FOR GLOBAL CNN AND WTBS

"I'm here to serve as the communicator who gets people together. I want to start dealing with issues like disarmament, pollution, soil erosion, population control, alternative energy sources."

READING TO BARBARA PYLE AT DINNER AT 1980 AMERICA'S CUP TRIALS IN NEWPORT

" 'The Global 2000 Report to the President.' The available evidence leaves no doubt that the world, including this nation, faces enormous, urgent and complex problems in the decades immediately ahead...

"Population in 2000 will grow from four billion in 1975 to six and one third billion by 2000. By 2100, population will grow to thirty billion, the maximum carrying capacity of the entire earth...

"Regional water shortages will become severe. There will be significant losses of world forests, serious deterioration of agricultural soils worldwide. Extinction of plant and animal species will increase dramatically. Hundreds of thousands of species – perhaps as much as twenty percent of all species on earth – will be irretrievably lost as their habitats vanish...

"If present trends continue, the world in 2000 will be more crowded, more polluted, less stable ecologically, and more vulnerable to disruption than the world we live in now. Serious stresses involving population, resources, and environment are clearly visible ahead...

"Barring revolutionary advances in technology, life for most people on earth will be more precarious in 2000 than it is now – unless the nations of the world act decisively to alter current trends."

EYEWITNESS

Barbara Pyle: "The report affirmed Ted's darkest fears about the future of the planet. He had been a closet environmentalist. He'd never had anyone to listen to his concerns. I became his sounding board. We had long talks and he offered me a job in Atlanta, to make documentary films for WTBS, but I told him I was not qualified. 'I don't care,' he said. 'Teach yourself to make movies the way you've taught yourself everything else.'"

THE ENVIRONMENTAL CONSCIENCE OF TURNER BROADCASTING

Ira Miskin: "If there was a person who educated Ted to the environment, it was Barbara and not the other way around. Despite what Barbara may say and despite what Ted believes, Barbara Pyle is the environmental conscience of Turner Broadcasting. Ted embraced those ideas, but Barbara has been talking about saving the environment since she could talk.

"They're like two crazy people. They're good friends. But they're both speeding freight trains at a hundred and fifty miles an hour all the time, no stopping, unrelenting. It's like you've got a brother and a sister who are exactly alike, and they have the same personality, and they can't be in a room at the same time for too long because they explode. That's what it is. It's a nutty brother-sister act."

ON RESPONSIBILITY FOR OUR SPECIES

From acceptance speech to cable industry on receiving Paul White Award for TV Excellence: "Kids today grow up in front of a television set. They don't read. I was a vociferous reader when I was young. I don't know whether your children read as much as you did, but my own five children sure didn't read as much as I did, and I did everything I could to get them to read and couldn't get them to read. So I can see that if we're going to educate the next generation of children, we're going to have to use video a great deal to do it.

"I really think we have a responsibility, because television news is so powerful, not to make a lot of money, but to have influence in our communities. And our community, as I interpret it, is not just the local market or even our country, but the world in which we live.

"And many of our problems are global and can only be solved together. I'm talking about problems of the environment, of population, of nuclear weapons and proliferation, of chemical and biological weapons – not to mention the poverty that exists in the world, the homelessness, and the hundred million children that are abandoned in the world today, while the world continues to spend a trillion dollars a year on armaments, enough money where even a half or a third of it, properly utilized, would provide decent health care and education, housing, food for everybody on earth that's living in abject poverty.

"We've got to change the way we're doing things, if we are going to survive. That's the conclusion I've come to, in studying the global situation very, very carefully, which I think, as a news person, we're sitting there with the information right at our fingertips, and we have to interpret it."

FOUNDING THE BETTER WORLD SOCIETY

"For the production of television programming on issues of critical importance to the survival of the planet."

"The Better World Society raises funding and commissions programming on what we feel are the critical issues of the times: the nuclear arms race, aiming toward peace on earth and controlling the population and preserving the environment.''

ON MAN AS AN ENDANGERED SPECIES

"We're doing a number of documentaries on the critical subjects. We believe that broadcasting has a responsibility. And maybe that's an old fashioned notion, but serving the public means something to me.''

"When you have critical issues like nuclear arms, ecological devastation, and a global population explosion putting man in the category of endangered species – that's Captain Cousteau and a lot of other futurists talking – broadcasting has a responsibility to point these things out to people. That's what Better World is trying to do.''

"The Society is obtaining funding from individuals and from foundations who are in the eleemosynary business and are producing and buying programming, not just for Turner Broadcasting. Our biggest series to date is going to run on public broadcasting. That's the ten hour series that Better World put $250,000 into. It was done by the BBC and Chinese television, and it's going to run on PBS this fall.''

JOHN DENVER INTRODUCES JACQUES COUSTEAU TO TURNER

Jacques Yves Cousteau: "It was love at fist sight. Ted had the same interest in ecology, in peace, in Third World problems – everything I'm interested in.''

ON MEETING ONE OF HIS CHILDHOOD HEROS

"God bless you, Captain Cousteau, you have been a great inspiration to me.''

COUSTEAU ASKS TURNER TO FUND HIS AMAZON EXPEDITION

TURNER: "How much money do you need?"

COUSTEAU: "Six million to fund the expedition and produce six or seven documentaries for TBS."

TURNER: "You've got it. Go to work."

"I gave him $4 million for his work this year. We'll get four hours of programming out of it. Of course, I'm losing my shirt on it. That's double the budget of network programs. But at least I'm keeping Cousteau operating. He's on my team." (WTBS supported Cousteau's expeditions for the next seven years.)

ABOARD CALYPSO ON AMAZON RIVER WITH TWO OF TURNER'S SONS

TURNER: "Do you think there's hope for us, Captain Cousteau?"

COUSTEAU: "Well, Ted, even if I knew we were going to lose, I would still work as hard as I could to save us. What else can a man of conscience do?"

COUSTEAU ON TURNER

Jacques Cousteau: "This man places his money and his actions where his heart is. So even if he is a little rough outside, he's the most sincere man I've ever met. I believe in what he says."

TURNER THROWS 75TH BIRTHDAY PARTY FOR JACQUES COUSTEAU AT MOUNT VERNON

"God bless you, Captain Cousteau." (John Denver, Cousteau Society board member, sings song he wrote, "Aye, Calypso.")

CASTRO, IMPRESSED WITH CNN'S EVENHANDED COVERAGE, INVITES TURNER TO CUBA

"I'm a very curious person. It was the first time I'd ever been to a communist country, and I was just interested in learning a little about how it worked. I just went down there as citizen Turner."

"After three drinks with rum, I got up my courage and I said, 'Are you interfering in Nicaragua and Angola?' Castro said, 'Yeah, you are, too.' And I said, 'Yeah, but we're the United States. We've got every right to be there.'

CASTRO: 'How come?'

TURNER: 'Because we're right. We're capitalists. We've got a free country.'

CASTRO: 'Yeah, but what about people that don't agree with that?'

"I went back and scratched my head. I never even thought there was another side to the picture."

"Castro's not a communist. He's like me – a dictator."

"I'm the only man on the planet ever to fly on Cuba's Air Force One with their president and on America's Air Force One with our president. (Pulls out photo album of his 1982 hunting trip with Fidel Castro)

"Look, this shows what I'm talking about. People are not all that different – all this killing and arms race is for nothing. Here's the great Commie dictator we're so worried about – having a hot toddy! Ha! And look, here's the great Commie dictator in his bare feet!

(Points to picture of Castro and Turner in camouflage outfits) "Here's us hunting. Twenty-two attempts on his life by the CIA – and I'm sitting next to him with a loaded rifle! Can you believe that?...

"I could've shot him in the back! He gave me the award for shooting more ducks than anyone he had ever hunted with. Good pictures? Great pictures!"

CALYPSO FOLLOWS COLUMBUS' 1492 ROUTE AND STOPS IN CUBA AT TURNER'S REQUEST

Jacques Cousteau: "The divers found a sea brimming with wealth... a rare abundance of fish in lush coral jungles.

(Surfacing from a dive himself) "Fantastique! I have never seen anything like it – a ballet of tarpon – everywhere."

NAUTICAL DIPLOMACY

With abundant charm, Cousteau requests that Castro free fifty political prisoners: (letter to an artist, Lazaro Jordana) "I am extremely happy to announce to you that you will soon be liberated, and to wish you good luck in your new life."

CASTRO FREES FIFTY POLITICAL PRISONERS

Lazaro Jordana: "Cousteau saved my life, and my father's life, and the lives of all the fifty people released. Then he didn't say anything about it. No publicity. I really like that about him. He's the kind of man who does things – doesn't talk about them – just does things."

TURNER CREATES "WORLD REPORT" AT CNN WITH 87 COUNTRIES PARTICIPATING

"After starting CNN itself, I think the best thing we have done is World Report. Basically, it's just giving people in the smaller countries, and countries from all over the world, a chance to be heard from."

"We've finally got them talking to each other instead of just fighting."

"As an American, I always thought of the world as 'President Reagan talked to Margaret Thatcher' or 'President Bush talks to Gorbachev,' but on 'World Report' you'll see the president of an African country meeting with another African leader. Americans never think that there's anything else going on in the world. The average American doesn't know from squat."

VISITING PYRAMIDS, MIDDLE EAST WEEKENDING AT PALACE OF KING HUSSEIN AND QUEEN NUR AND (ONE EVENING) WITH YASSIR ARAFAT

"There needs to be peace here. I want to bring peace. I want to bring these countries together."

VISITING PANDAS, CLIMBING GREAT WALL OF CHINA, AND SIPPING GREEN TEA WITH PREMIER LI-PENG

Kathy Leach: "For the next half hour they talked of the environment, of population control, and – hopefully – of the world situation."

CNN BYPASSES BARBED WIRE

Joshua Meyrowitz, professor: "Many of the things that define national sovereignty are fading. National sovereignty wasn't based only on power and barbed wire; it was based also on information control. Nations are losing control over informational borders because of CNN."

SECRET KGB COUP D'ETAT ON GORBACHEV FAILS AS WORLD WATCHES AND INTERVENES

Mikhail Gorbachev (In upstairs office at summer dracha on Black Sea): "Before I could invite them in they had already gone into my office. I wasn't used to such conduct, such familiarity. They were led by the head of my presidential staff, Valeri Boldin..

(Told to sign or resign, yells) "You are nothing but adventurists and traitors and you will pay for this. Only those who want to commit suicide can now suggest a totalitarian regime in the country. You are pushing it to civil war......

(Again told to resign) "You'll never live that long!"

Raisa Gorbachev: "We have a military hot line in each of our homes. That phone was kept under a cover. No one is allowed to touch it, not even to dust it. Mikhail took off the cover, lifted the handset, and it was dead. Then we knew. That was that. When we realized we were under arrest, and when we discovered who the conspirators were, we were terribly hurt. We felt bitterly betrayed." (It turned out the plotters had also taken Gorbachev's nuclear briefcase.)

EYEWITNESSES

Steven Hurst, CNN anchor (reporting the next day with camera crew from the roof of the Russian White House): "Boris Yeltsin played the hero. Clambered aboard a tank – bellowed about unconstitutionality. Demanded Mikhail Gorbachev's return from house arrest. The few hundred assembled, cheered. Tank drivers, heads wrapped in leather caps, popped out of armored hatches, bewildered. The camera atop our building was still live to the world. The vigil began.''

HOMES AROUND MOSCOW COULD RECEIVE THE NEWLY INSTALLED CNN WITH RABBIT EARS

Boris Yeltsin (on top of tank): "Citizens of Russia... the legally elected president of the Country was removed from power... We are dealing with a rightist reactionary unconstitutional coup..." (1:25 p.m.: Russian TV is forced off the air. CNN continues live to 123 countries)

Steven Hurst: "Inside the White House defenders had ordered women and children to leave the vicinity. Gas masks and bulletproof vests were passed out. Someone started a rumor that the White House had been mined against an attack. I got frightened. My wife, Claire Shipman, was in that building, refusing to leave. She was watching when violence finally broke out, and reporting live for CNN as defenders attacked a tank with Molotov cocktails."

Claire Shipman, reporting for CNN from inside the White House with a phone and camera crew on the balcony): "From this vantage point, the mesh, metal and sticks below appear a formidable fortress, or at least a hurricane effort.''

NEXT DAY

Steven Hurst: "Day Two. Fatigue joined depression. The rain fell in earnest. And amazingly, none of the men running the putsch had yet got around to shutting us down. Why was our camera – bouncing its damning images off a Soviet satellite – still live from its immodest vantage point, glaring at this stand-off between Communist armor and a rag-tag assemblage of the unarmed and mainly curious?

"The camera on the roof, its images fitful because the rain was so strong, electrical connections disturbed, showed the crowd expanding at Yeltsin's headquarters."

Boris Yeltsin: "There was an elite group of the KGB whose orders were to destroy the first two floors of the building – machine gun everything. Their main task was to seize me – and, if I escaped – to shoot me."

Claire Shipman: "Defenders have locked arms, barricading the building themselves. The monotone voice on the intercom informs that the attack is now expected in two hours. We commandeer a phone – it becomes our lifeline to CNN in Atlanta. The evening becomes a series of trips between the Yeltsin information center (CNN had provided a TV so they could monitor events and reaction of world leaders deploring the coup) and our camera and phone line on the balcony... We get a care package from our bureau: food, bulletproof vests, other gear. Outside we see the crowd has thinned, but thousands remain despite the curfew and the rain. Candles flicker in the hands of solemn onlookers as a Russian Orthodox priest holds an extemporaneous service. It's eerily reminiscent of Tianenmen Square."

Neil Felshman: "In the West the story on the streets of Moscow was played out, blow by blow, on CNN. That story got back to Moscow and affected events there. The communications technology that was only beginning to be available in the Soviet Union – plus the twenty-four hour-a-day seven days a week coverage of news events throughout a global communications community – made control of information needed for a successful takeover of government impossible."

TWO KGB LED PLANES FLY TO CRIMEAN COMPOUND

A delegate from Russian Parliament group flying to rescue Gorbachev from his Black Sea dracha: "The plotters of the coup have flown secretly to Gorbachev. They've misled us. They've tricked us all. We are afraid of what might happen to Gorbachev."

DUE TO HUGE CROWDS AND WORLD RESISTANCE, COUP COLLAPSES

Defense Minister Yazov, under balcony at dracha: "I wanted to sink into the ground. I felt eternally guilty for what we had done to Gorbachev and Raisa. I felt guilty in front of the people and the party."

(Mikhail Gorbachev later thanked Ted Turner in person for probably saving his life and preventing a return of the U.S.S.R. to hard-line dictatorship.)

AVALANCHE OF REFORM MOVEMENT, MAGNIFIED BY CNN, SWEPT AWAY THE U.S.S.R. WITHIN MONTHS

Boris Yeltsin: "Gorbachev started by climbing a mountain whose summit is not even visible. It is somewhere up in the clouds and no one knows how the ascent will end. Will we be swept away by an avalanche? Or will this Everest be conquered?"

PRESS

Time: "The motherland of communism overthrew its leaders and their doctrine - before the cameras.

"To a considerable degree, especially in Moscow, momentous things happened precisely BECAUSE they were being seen AS they happened. These shots heard – and seen – around the world appeared under the aegis of the first global TV news company, Cable Network News.''

ON GOALS FOR FUTURE

"Everyday we reach more people as more satellite dishes and cable systems hook up in the United States and all over the world – and the more people we reach, the more influential we'll be. We're a long way from peaking in terms of influence.

"But, there's no question in my mind that we've been a force for moderation and intelligence in the world. We have interviewed world leaders on a global stage for the first time. And every leader wants to be favorably thought of. Everyone wants to be popular. Even Adolph Hitler wanted to be popular. He thought he was doing the right thing...

"But there is a growing consensus of what the right thing is. And the right thing is peace and moderation and human rights, and you see it popping up all over the place – even in South Africa, in the demise of the Berlin Wall and the freing up of Eastern Europe, in Vietnam pulling out of Cambodia."

THE GLOBALIZATION OF THE GLOBE

"I mean, it's just good news – the rising environmental and population concerns all over the world – and it almost makes you think that humanity has a chance. Some old sage said years ago that a problem recognized is a problem half-solved. And I really do think that is true.

"You have got to basically believe that human beings, when confronted with the information, will choose the intelligent course. If they don't, then there really was no hope for them in the beginning....

"With the right information, we hopefully are going to make the right decisions."

SETTING UP CNN BUREAUS IN EUROPE AMIDST RESISTANCE

"I don't want to upset the apple cart or make anybody mad. I just want to be a banana, one of the bunch."

"I think what has occurred has been pretty terrific. In less than eight years we've become the first global network. And we're highly respected by virtually everyone. We don't get criticism from the eastern bloc; we don't get criticism from the conservatives or the liberals. We don't get complaints from the Ayatollah even; we had the Iranian ambassador to the UN talking."

TRAVELING, WIDENING, BROADENING

"I've already met or exceeded all goals for personal wealth and accomplishment. Nothing ever came easy – my first eight years of sailing I didn't even win my club championship. But I just kept working and working and working, that's the secret of my success. Now I am like a runner who has kept running and running – and one day finds he has run the Boston Marathon. I don't need to be the best anymore, I'm just part of a team, I'm just widening, broadening."

"I wanted to use communications as a positive force, to tie the world together, and – you know something? It's working!"

CHAPTER TEN
SHAKING UP THE NETWORKS

TO GROUP OF ADVERTISERS

"I've got everything I need personally. I've got a baseball team, a basketball team, a soccer team, two sixty-foot yachts, a plantation, a private island, a farm, a wife, five kids, and two networks. Ain't never been anybody in the history of the world ever had more than me. So let me tell you what I'm going for. I'm going now for the history books. I'm swinging for the fences. And I don't want to be remembered as a bad guy.

"And, since I'm in the advertising business, I want to be the hero of the advertising business. I'm making these promises to you. I'm going to deliver like I've delivered in the past. And I'm saying we're going to be in 50% of the homes in this country on January 1, 1985, with both networks. We're wiring the whole damn country, and we're going to do it together. We're going to make a ton of money. We're going to do a lot of good. And we're going to have a lot of fun.

"Intellectually and morally the (Big Three) networks are tearing this country down. Television is the most powerful form of propaganda the world has ever seen, and in our nation it's being used to destroy us."

ON ROLLING THE DICE WITH
HIS ENTIRE FORTUNE

"It doesn't bother me that I'm committing almost all I have. Had I known I was going to fail when I started, I would still have done it because it needs to be done - Of course, I also think we'll make a fortune."

"Sure, I'm worried. But I'm not that worried. As soon as I earn me my billion dollars, I am going to buy a network. I am going to find the new Frank Capra and set him to making movies. I can quit whenever I want to. I am not worried about what people think. But I am the right person in the right place at the right time, not me alone, but all the people who think the world can be brought together by communications."

"We'll have enough homes on the news networks this fall where we'll be breaking even. We're not that far from breaking even. Between now and the end of the year, the news network's coverage will double...

"By January 1, 1985 we'll be in half the homes in the country."

"My biggest job is in Washington. The entire entrenched entertainment and telecommunications business wants to see me destroyed because I pose a terrible threat to their whole rotten set up. So I've got to protect myself first in Washington. I need the cable industry to carry my programs... I mean, believe it or not, as good as Cable News is, over half the cable industry is not carrying it. The most important thing is selling the advertisers and the advertising agencies. But if I get the ratings, if I get more homes, the advertisers have got to come along."

CNN WAS NOT ALLOWED TO COVER WHITE HOUSE NEWS

TURNER: "Cable News Network today files suit against ABC, CBS and NBC seeking damages... and relief for violations of the anti-trust laws of the United States, resulting from their practice of pooling news coverage.

"Also named in the suit are President Ronald Reagan, White House Chief of Staff James A. Baker, and Deputy Press Secretary Larry Speakes, for violation of Cable News Network's right to equal access to cover presidential news conferences, trips, and other presidential activities and White House events, as guaranteed by the Constitution of the United States.

"The three major networks should not be able to dictate the terms of our coverage, limit its scope, or combine their resources to effectively prevent CNN from competing with them, and the president's staff should not be allowed to deny CNN equal access to cover presidential activities and White House events."

SCHONFELD: "If the White House has the right to name who is going to be a pool, it becomes a fundamental question of freedom of speech. We believe the networks are pressuring the White House to keep us out. We can compete with any of the three networks one-on-one. It's when the three of them gang up that we have problems."

TO PRESS

"I demand an immediate Congressional investigation to determine whether (network programming) has a detrimental effect on the morals, attitudes, and habits of the people of this country...

"The networks are worse than witches. They need to be hunted down and prosecuted."

BACK ON HIS WHITE HORSE

"What those networks are doing is making Hitler youth out of the American people – lazy, drug addicts, homosexuals, sex maniacs, materialists, disrespectful. I mean, you know – mockery of their parents, mockery of all the institutions. It's bad. Bad, bad, bad. They ought to be tried for treason; they're the worst enemies America ever had."

"We're going to do a bunch of investigative exposes on the networks. Listen, I'm going after the networks. I'm going to scare the Hell out of them."

"We are at war with everybody. Now we have lawsuits against the three networks, the White House and Westinghouse. And we're winning – at least we haven't lost yet."

TO 7000 VETERANS OF FOREIGN WARS

"The worst enemies the United States ever faced are not the Nazi's and the Japanese in World War II, but are living among us today and running the three networks."

CRUSADING WITH LIVE EDITORIAL ON CNN

"I would first like to point out and make very clear that I am against news censorship of any type. But after watching the last few days the trial of John Hinkley, Jr., for the attempted presidential assassination of Ronald Reagan, I am very, very concerned that this movie, "Taxi Driver," was an inspiration to Hinkley and was partially to blame for his attempted assassination of our president. Many years ago I stumbled on that movie.... and was absolutely appalled by the blood and gore, and I knew, at that time, that movie was going to have an effect on young people...

SHAKING UP THE NETWORKS

"The people responsible for this movie should be just as much on trial as John Hinkley himself...

"Write your Congressman and your Senator right away, and tell him that you want something done."

"The Big Three are an evil empire. They're in desperate trouble. The networks are like dinosaurs. Most of their money is going to the outrageous costs of Hollywood productions, going to pay for cocaine, so your favorite TV people can do cocaine...

THE NETWORKS ARE LIKE DINOSAURS

"You know why they're not here anymore? Because the mammals ate their eggs. I'm a mammal."

"There are a lot of things in broadcasting that don't make sense. It doesn't make sense that three networks got their owned and operated stations for free forty years go, that they paid no spectrum fee and that they continue to get those licences renewed. It's free use of a public resource by a private company. No one would ever think of allowing Exxon or Texaco to drill on off-shore land without having bids like they do.

"If something doesn't make sense, it's the use by the broadcasters of the spectrum without paying anything for it. If they're going to go back and straighten everything out they ought to put up all the frequencies for bids. They'd clear up the national debt."

LOOKING BACK

"We've created havoc for the big three networks over the years. All three of them had to change hands. How much more havoc can you have than that?

"Only one of the networks is really profitable. And with cable in 50% of the homes, I don't think the future of the three networks is nearly as bright as it was – the future has already caught up with them."

"You tell me what's going to improve their basic underlying values."

AS LATE AS 1987, CABLE STILL MEETS WITH POLITICAL RESISTANCE IN NEW YORK CITY AND WASHINGTON

"What John Malone (TBS board member, cable operator) said about finishing the wiring of the metropolitan areas that are currently not wired, is key. The boroughs of New York, the District of Columbia, and downtown Cleveland – these things are happening slowly, but they are finally happening. That's important because having cable available to everybody is going to make it more politically acceptable for major events to migrate to cable if everyone has the availability of it. To be really, truly a national medium, cable has to be available everywhere.

"It's important that the packaging and the scrambling be made easy for people with earth satellite stations, and it's getting easier to get the descramblers."

"The network ratings this summer are down 15% over last year. That's catastrophic to go down that much. Last year they were able to say that erosion stopped. That's like somebody who's got cancer getting some of those radiation treatments and it looks like for a month that the cancer has been arrested, and then they check and there's ten more spots breaking out all over their bodies."

VIOLENCE AND SHABBY MORAL VALUES ARE NOT THE PEOPLE'S CHOICE

"You take the value of the cable industry today – a one hundred billion dollar industry – and all the broadcasters together would be $40 billion. And the cash flow is there too, because the cable industry is doing thirteen million dollars and it's making 50%. It's got a six billion or seven billion dollar cash flow. The three networks together only have a three hundred million dollar cash flow. The whole broadcast cash flow wouldn't be three billion dollars.

"The cable operators are going to spend more on programming mainly because they're running out of other places to put the money. That's a great situation – to be awash in an industry that's just rolling in it."

TURNER IN COWBOY HAT ON BILLBOARDS, "I WAS CABLE WHEN CABLE WASN'T COOL"

(Goes to recording studio to tape his part in promotional tape):

"Down in Atlanta, Georgia

(the musician was strummin' and singin'),

There's a legendary dude.

In New York they say he's crazy
In Newport they call him rude
And to those Bel Air mansions
Where Cadillacs all go
He never gets invited
'Cause he's just a good ol' boy.

Though society may shun him
'Cause he won't toe their line

To the people across the country
Teddy Turner's doing fine.
He was cable when cable wasn't cool.

When he went cable
They called that boy a fool.
The experts said his time ain't come
And in the end he's number one."

"Heck, all I did was jump the gun." (Ted's line)

"In those network ivory towers
They still look down their nose
'How can anybody start a network
Down where the cotton grows?

You gotta be near Broadway
Or at Sunset and Vine.
You gotta be sophisticated
To program that prime time.'
But this is their perspective
They're bogged down in the past.
And NBC, CBS and ABC won't last.

He was cable when cable wasn't cool.
The Captain Outrageous
He's broken every rule."

CHAPTER ELEVEN
THE CABLE NEWS WAR, CBS BID AND MGM

"Everything I do is a war. It's a war between the forces of good and evil, hatred and stupidity, greed and materialism versus the forces of light."

WITH MUCH ZEAL BUT SHOESTRING BUDGET, CNN STAFF RESORTS TO BEGGING ADVANCE PAYMENT FROM ADVERTISERS AND BRAVES PEANUT VENDORS TO MAKE WEEKLY PAYROLL

Reese Schonfeld: "We'd take ninety percent of the receipts due, if we could get them NOW, to get us through – it was that hand to mouth."

SENSING WEAKNESS, ABC/WESTINGHOUSE CHOSE THIS MOMENT TO ANNOUNCE THEIR OWN 24 HOUR NEWS CHANNEL, A HEADLINE NEWS VERSION

(Slicing the air in the CNN offices with his broadsword) "It's a pre-emptive first strike! There is just one response – attack!"

WALKING THROUGH CNN OFFICES, STILL BRANDISHING SWORD

"It's an immediate counter-attack – just like Franklin Delano Roosevelt's after Pearl Harbour. When the Japanese carried out their surprise attack on the Hawaiian base Roosevelt didn't hesitate; he immediately declared war and went on the offensive. And that's exactly what we're going to do. We're going to offer everything they say they're going to offer, except ours will be on the air first and it will be BETTER! They only have money to lose, but we have EVERYTHING to lose. We have to stake out this territory now or we'll lose it forever. This is the time in cable history where you either establish your bona fides or you don't!" (Pulls the plan Reese Schonfeld had already written for Headline News out of his desk drawer, and puts Kavanau in charge of launching it.)

EYEWITNESS

Reese Schonfeld: "Ted's happy. We're at war again."

TED HAD SPENT THE WEEKEND ON HOPE PLANTATION WALKING ALONG THE EDISTO RIVER THINKING HARD

"It was basically a defense, but in defending you often attack. We were holding the ground. We had CNN already and they were going to attack with two channels...

"I figured – knowing they were two big companies and that they were both public corporations, and how slow those kinds of operations usually run – that if their losses were bigger than anticipated, the people in charge of this project would come under criticism...And I knew that if they ran into unanticipated difficulties there would be friction between the two fifty-fifty partners...

"So even though we were very, very strapped financially, and they knew it, I decided that we would beat them to the market. We would split the market for that service, so they would not be as viable. I didn't know exactly how long we could last. I think the two of them had resources a hundred times greater than mine...

"I did know that in a war of attrition we would lose."

ON CNN'S "TAKE TWO" SHOW

FARMER: "What was your first reaction when you heard that Westinghouse and ABC were going to start another cable news service?"

TURNER: "Well, Don, it's real INTERESTING, because all the networks pooh-poohed the idea when we first announced we were going to do it.

"Now they're coming in because they can see that we're becoming tremendously successful. Westinghouse is the largest group broadcaster in the nation outside of the three networks. But in Washington today, the Reagan administration is letting anything go...

"With ABC's capitalization of $800 million, we now have a raid against us, committed to our destruction, of approximately $4 billion – versus our capitalization of $200 million."

CURLE: "You're in a remarkable good humor today, Mr. Turner."

TURNER: "They're fifty times bigger but with the Superstation and the Cable News Network, we've always been the little guys fighting the big guys, and I really relish the fight."

FARMER: "Can you keep going? There are a lot of rumors around in financial circles that Ted Turner, no matter how good his intentions, no matter how good his product – is going to need the help of some OTHER conglomerate to FIGHT the conglomerates."

TURNER: "I will do whatever is necessary to survive... THE ONLY WAY THEY'RE GOING TO GET RID OF ME IS TO PUT A BULLET IN ME..."

"I really like our company being independent and I hope we can stay that way."

"TAKE TWO" DOES POLL, RESULTS 85% CONSIDERED CNN EXCELLENT AS OPPOSED TO GOOD OR FAIR

TURNER: "Holy Smoke! You know, when we started this whole thing – because I knew we were walking in where, in fact, one of the articles said, angels fear to tread – I said it then and I say it now, that even if I knew we were going to fail when we started I would have gone ahead and done it anyway, because even if we were only on for a short period of time we would have shown what responsible journalism on television could do."

ANNUAL CABLE SYSTEMS OPERATOR'S CONFERENCE

"When I cast my lot with the cable industry ten years ago, I DREAMED that all these things would happen.."

"And – it's not unexpected that the three guys – the second wave companies, the three major networks, would be in a state of panic and try to figure out a way to cut themselves in and get control of the news on your cable systems, at a belated time, when they see what tremendous damage we are doing to them!

"You know, I've always been a fighter. And I remember when this industry was scrapping for survival. We've always been scrapping. We've had to fight for our right to live against these networks that are now standing in line to utilize your channels. We had to fight for our lives. You'll find that I've spent a great deal of time and money

THE CABLE NEWS WAR, CBS BID AND MGM

in Washington helping you fight these battles – because your battles have been my battles ever since I joined this industry..."

"Timing is everything and the good Lord and destiny are looking after both of us..."

"The one thing is, you guys and girls and ladies in the cable industry, you are learning. A year ago, you knew very little about news, but you're learning a lot about it. I mean, I didn't know much about it either. But I knew enough to know that the broadcast news approach was the wrong way to do it..."

"They've turned our people against their wonderful government, and they've turned our people against business...they're anti-religion, anti-family, anti-American!"

ATTACKING PLANNED ABC NEWS CHANNEL AT SAME CONFERENCE

"Anybody who goes with them is going with a second-rate horse shit operation! Any questions?"

TO CNN STAFF

"I'm really proud of what we've done and I know everybody here is, too, and we've worked hard, haven't we? I mean, this has been the hardest working group of people that the world has ever seen! We're going to keep in there swinging. You know, onward and upward. I don't know where we're going, but we're going there in a hurry. Awwriight!"

AFTER CNN SUED ABC, SIGNED UP THEIR CABLE SUBSCRIBERS, AND PAID ABC $20 MILLION

ABC signing off: "And now, Ted buddy, it's all yours."

EYEWITNESS

Ira Miskin: "I don't know how close he was to throwing in the towel near the end of the CNN-SNC battle, but we were pretty broke, and pretty desperate at that point. I think more his bravado than his maneuvering got the suits at ABC and Westinghouse to flinch. Because, in truth, they had the pockets; they could have beaten us. They could have been the network. And I think he was very lucky that they believed that he would go until the last drop of blood."

EYEWITNESS

Mike Gearon: "How much of it was luck? To what extent was all this destined to happen? When you go back and examine it, there had to be a certain amount of fortunate timing. But how much of it was just because of this guy's tremendous will? Ted's successful because he's just so relentless. Ted's a guy that if you went to war, you'd want him on your side. Tomorrow I'd venture if you were to scratch the great military generals and redo World War II, he'd be up there with the all-time greats. Because he knows how to go to battle. And he knows how to win. Was Patton a great salesman? I don't know if he was a great salesman or not but he damn sure was going to scare the shit out of you and convince you that he was going to win."

CNN NOW CLAIMED THE CABLE NEWS TERRITORY

"On a level playing field, ABC and Group W got their brains kicked out!"

"We might see a little black ink as early as November, that is, if Schonfeld doesn't come up with some riot in Europe that costs millions of dollars to cover. Even so – we'll be looking real good in 1982."

THE CABLE NEWS WAR, CBS BID AND MGM

NEW CNN CENTER, A HOTEL, A COMPLEX, SIX THEATERS

"The office building and retail space was only twenty-five percent rented when we bought it a year ago, and now it's about eighty percent rented. But basically we took most of the space ourselves. The hardest space to do anything with, the old amusement park space, was just perfect for our new facilities for CNN. We bought the place for sixty million; it would cost a hundred and eighty to reproduce it today. It cost a hundred and twenty million ten years ago.

"We're one hundred percent financed by the Canadian Imperial Bank, which had taken it in bankruptcy. Their officers were there last night (at the official opening of CNN Center) and just about gave me a hug. They lent us, I think, another ten million to remodel the hotel and do the other things that we need, so we've got nothing in it. I think some Japanese companies were talking about giving us a hundred and seventy-five million for it, so we have a one hundred million dollar profit in it in less than a year. We've turned it from a bankrupt operation into a very viable one."

CBS GOES TO ATLANTA TO TRY TO BUY CNN

TURNER: "I'll sell you CNN. How much of it do you want to buy?"

CBS EXECUTIVES: "Fifty-one percent or more."

TURNER: "You want control? You don't buy control of Ted Turner's companies. Forty-nine percent or less."

CBS: "No, honestly, Ted. You know CBS. We just wouldn't be interested in less than fifty-one percent..." (Talks continued for two and one half hours.)

TURNER: "Forget about buying CNN. Let's talk about CBS merging with TBS. We got the greatest news organization in the world. We're non-union and we're going to open bureaus all around the world. I can help you. We can team up and we can both cut costs. We don't both need bureaus in every place.

(CBS Executives head back to plane.) "You guys, you CBS guys are something. Someday I'm going to own you. You bet I am. Remember I told you so."

TURNER WITH CBS IN NEW YORK

CBS: "I'm here to buy you."

TURNER: "I'll buy you."

CBS: "That's ridiculous. I'm buying you."

TURNER: "No, I'll buy you."

CBS: "We'll get back in touch with you."

AT WASHINGTON PRESS CLUB

"CBS came down to Atlanta trying to acquire CNN. I didn't let them make an offer. I was too busy telling them I was going to bust them."

RETURN VOLLEY FROM CBS

Thomas Wyman: "We were invited to come to Atlanta by Mr. Turner. He said he wanted to talk about some sort of collaboration...

"But the suggestion that we were dying to make an offer for CNN is outrageous and untrue."

TURNER SERIOUSLY GOES AFTER CBS

"As far as CNN was concerned, trying to take over CBS made sense from the standpoint that you could consolidate your national and international newsgathering facilities. You could gather the news for CBS and CNN with one organization. That's $100 million in savings

THE CABLE NEWS WAR, CBS BID AND MGM

right there, if you wanted to save. Instead of two bureaus in Moscow, say, you could have just one. But if you wanted to keep spending the same amount of money, you could have twice as big a newsgathering organization. Instead of having twenty overseas bureaus, you could have thirty-five. You could have bureaus virtually everywhere in the world.''

"I needed distribution
and CBS represented distribution."

EYEWITNESS

Robert Wussler: "Ted's father imbued in him the spirit of quality. The result was that, like his father he saw CBS as terrific TV, National Geographic as the best magazine, MGM as having terrific musical movies. These things stuck with him."

IF TURNER BOUGHT CBS

"It's the Tiffany network, but also all that's wrong with television. I'd make changes in the entertainment programming like less violence, more pro-family, pro-American-type programming, and less sex and violence and stupidity."

"I probably won't make it. It's a long shot, but I'm going to take my shot. Even if the bid doesn't work, there will be other benefits. The publicity would make it easier to sell ads for TBS. We would be perceived as a more major player on Madison Avenue."

CAN A MINNOW SWALLOW A WHALE?

(TBS revenues were $282 million; CBS, $4.9 billion) "We have been very interested in joining forces with one of the three networks because we intend to be number one in the business.

(Offers $5.41 billion in stocks and junk bonds to CBS shareholders, a paper deal) "I'll never call them junk bonds. That's what the doomsayers call them. I like to call them high yield."

AT METROPOLITAN MUSEUM OF ART PARTY

(Surrounded by fans) "I don't know what all this ranting and raving is about. What am I but a free enterprise guy? The same way as Bill Paley when he started CBS. Paley was an outsider when he started, too.

(Kissing hands of socialites) "As I've grown older I've become less desperate. I've decided to be more statesmanlike. I'm going to be a gentleman...

"Gee, will you look at that? That's Walter Cronkite!..(Rushes over)

"You and I would get on like a house on fire! (Offers Cronkite a million dollars if he'd go back on the air)... When I own CBS..."

CBS TAKES POISON PILL - HUGE DEBT, BY BUYING BACK 21% OF ITS OWN STOCK

"Poison pill. The last guy I knew who took a poison pill was Hermann Goring."

EYEWITNESS

Mike Gearon: "In the end it was worth every penny Ted spent ($18.6 million). It really left CBS crippled in many ways ($954 million debt)."

BUILDING A CASTLE OF SAND OR A PYRAMID OF STONE?

"We were pretty solid in news – we were controlling our own destiny – because we were originating our own programming. Our news operations were as solid as the Rock of Gibraltar.

"We did not have any proprietary programming for WTBS to speak of. We had the Braves, and we had a little Cousteau and some of the documentaries that we'd done, but we were at the mercy of Hollywood.

"And I wanted to use our position to rise to the next level, which was to own a library – a fine library – so that we could control enough programming that we could program a really fine entertainment network, not only in the United States but globally, to position ourselves in the global business."

AS CBS TAKEOVER FAILS, TURNER GOES AFTER MGM STUDIOS

With MGM/UA's 4600 films Turner had stars in his eyes: "This is the answer to my dreams."

TBS BOARD MEETING

"To be a long term factor in this business you have to either grow in viewership (CBS deal) or in the programs you own...

"It's essential to have this additional programming to make TBS competitive. We've got to do this to survive. We NEED to do this...

"Movie fees are rising. Every time we sign a new contract (to rent movies) it's more, more, more..

"We NEED to do this. Let's get on with it."

"It's a business where the big are getting bigger and the small are disappearing. I want to be one of the survivors. This makes us more of a major player. It gets people's attention. I mean, we are in show biz..

"Without MGM there's a question mark with survivors about our long-term viability. They used to throw that up to me. Early on, there was a certain group of people in the New York establishment – the same sort of thinking that gave me such a hard time at the New York Yacht Club. You know: 'He's from Georgia, we don't need him around here. Here's a quarter, go shine my shoes.' You know: Step 'n' Fetchit."

BUYS MGM/UA FOR $1.5 BILLION, FULL ASKING PRICE

"I paid full value and the value to me was greater than the value to anyone else because I had a distribution system that wants old movies – and I like old movies. I've never done anything like this before. It's like sailboat racing in a hurricane. It's like being in an airplane in a storm. You buckle your seatbelt and hope for the best."

"Mr. Kerkorian is no dummy. He knew what he wanted to sell, and what he sold was the troubled part of the company...

"But I have always bought troubled things. Normally, things aren't for sale if they're in great shape. Right?"

"I knew we were in trouble before I bought MGM. That's why I bought it. I maintain, we were in much more trouble before we got into it"

"There were two ways to go. One was to get stronger in distribution, and that was the strategic reason for the move to acquire CBS. Or I had to get stronger buying power – if you're a strong enough buyer and have huge leverage you can basically force buying. We were not really strong in either distribution or program ownership, or not strong enough, and I didn't like our strategic position..."

THE CABLE NEWS WAR, CBS BID AND MGM

"I don't think this is the time to be small in the networking business. Even in syndication you see consolidation going on. I think that, long term, there are just going to be a few major companies in the programming and distribution business.

"I owe nearly $2 billion. That's more than the debt of some Third World countries. I'm pretty proud of that. Today, it's not how much you earn, but how much you owe."

"How can you go broke buying the Rembrandts of the programming business, when you are a programmer?"

IN PRIVATE

"Goddamn it! I've really done it this time. I may have really done it. Maybe I shouldn't have gone into MGM."

WINGING IT AT STAR-STUDDED HOLLYWOOD EVENT

Eyewitness: "In three minutes, Turner had the entire audience eating out of his hand. He had the entire INDUSTRY eating out of his hand. I don't think even one person in the room knew what he said. He just absolutely won everybody over with his incredible charm. This guy is incredible, really incredible!"

TURNER BECOMES LARGER - AND A BIT LOUDER - THAN LIFE

"Everybody wants to become a star. The garbage collector wants to be O.J. Simpson, didn't you know that? I decided to go with the mustache when I saw Rhett Butler in 'Gone with the Wind.'... 'Frankly, my dear, I don't give a damn.' What a line. What a time. You know where I'm really big? Australia. They really know their sailing there. Look at this (scrapbook). Huge coverage. All over the front pages. Damn. You talk about a star.''

"Dee, a taxi driver in New York recognized me! Can you believe that?"

WITH MOTHER

TURNER (Driving in car with some other people): "Two billion dollars I owe! Actually, it's closer to one point nine but I like the sound of two billion better."

MOTHER, from back seat: "Oh, my. How much did he say?"

TURNER: "TWO BILLION. No individual in history has ever owed more."

MOTHER: "Oh, Ted, I get a headache thinking about it. Well, you're honest. You'll try to pay it all back, I know."

TURNER: "That's a million dollars a day in interest, Mother. Here, look at my picture in today's newspaper. Do I look worried?"

MOTHER: "Oh, Ted. You're so full of the dickens. I just wish you had time to come to my house for dinner. It seems like the only time I talk to you any more is in a car, driving from one place to another."

COULD HIS WHOLE EMPIRE COME CRASHING DOWN AROUND HIS EARS?

"If a major recession caught me in a high-debt cycle, I'd be in trouble, but so would Chase Manhattan and a lot of others. I think my wife would stick with me. My dog would still love me.

"If I could afford birdseed, my cockatoo would stick around – a lot of friends wouldn't."

THE CABLE NEWS WAR, CBS BID AND MGM PUTS CNN ON THE BLOCK

(Time, Inc., Rupert Murdoch, USA Today, NBC are all in the running) "NBC gave me until 5 p.m. New York time to accept or reject their final bid. That's the kind of ultimatum Hitler gave the poles."

TURNER NIXES NBC OFFER

"I feel like Scarlett O'Hara. I'm down there digging in the dirt and eating radishes because I've got nothing else to do."

EYEWITNESS

Robert Wussler, on offer from USA Today: "That deal was the road to easy street...

(When Turner nixes that offer, too) "Ted is antiprint. He doesn't believe in the future of publications. He said he thought that newspapers would be dead in ten years. Ted's ecological sense was against newsprint, something you read and throw away every day. He always talked about that."

CHOOSES TO SELL TO A GROUP OF CABLE OPERATORS

"We had oodles of alternatives, and we picked the one that made the most strategic sense for us, and that was placing it in hands that were widely spread throughout the cable industry, rather than having one or two big partners."

THE 'BEST AND BRIGHTEST' CABLE OWNERS NOW OWNED 49% OF TURNER BROADCASTING

"I don't consider it a rescue at all. There were several other entities that were sitting there willing to do the same thing, maybe even on more favorable terms. We were not desperate in any way, shape or form.

"We just kept our nose to the grindstone and went quietly about our business. You didn't see any panicking on the part of our company. We didn't do like CBS did, with its massive layoffs in the news departments."

IN PUBLIC

"Life is a series of gambles, and you know the final outcome is going to be a tragic one. So since you're dead – the day you're born you start to die – you have to be willing to accept risks in business, and there have been risks all along. I think we're in a less risky position today than we've ever been.

"I have over 50% of the company. I don't even know what percentage of the voting stock I have – 65% of the votes? I do have some negative covenants, but the negative covenants I have with the cable operators are not significantly different than they are with the banks.

"We create more ink, even though we're not saying anything. And everybody says, 'Well, he's really done it this time!'

"I like to think that we're like Brer Rabbit. When he got caught in the briar patch, he was hoping he'd get thrown in that briar patch."

"They bought stock at 15, and it's now at 25 three months later. So they made 80% on their investment in three months. It's not dog meat by any stretch of the imagination."

IN PRIVATE

"I've lost control... I've lost control."

(Later) "At the time the MGM deal was a mistake. I mean if you say I just made a deal that required me to give away half my company, when that wasn't the intent of the deal.

THE CABLE NEWS WAR, CBS BID AND MGM

"Well, it might work out in the future, but at the time it was clearly a miscalculation. They totally miscalculated the deal."

BOARD GIVES TURNER $2 MILLION SPENDING LIMIT

(Now has 14 cable operators on his board of directors and holds only 51% of the stock): "I won't be able to make any large acquisitions or do any large projects without board approval. My power will be somewhat diminished."

TURNER'S 1992 NEW YEAR'S RESOLUTION

"Peace on Earth
Goodwill to Men
To Get my Company back"

BEGINS COLORIZING CLASSIC MOVIES

"It's obvious I was seeing things that other people didn't see, just like Columbus discovered the world was round. Women put on make-up, don't they? That's coloring isn't it? Nothing wrong with that. Besides, when was the last time anyone took photos in black and white – I know, Ansel Adams. But he's dead, too."

HOLLYWOOD PURISTS DEPLORE COLORIZING

Newsweek: "Turner's reputation in Hollywood may already be beyond repair. There had been talk of a new breed of Hollywood mogul, a visionary, a showman. But since his purchase of MGM, Turner's image in Hollywood has gone from swashbuckling new man in town to the Phillistine from the South."

John Huston, director, "The Maltese Falcon": "It is as great an impertinence as for someone to wash flesh tones over a da Vinci drawing."

Woody Allen: "a criminal mutilation"

Jimmy Stewart: "cultural butchery"

TURNER ALSO ACQUIRED TWO DOZEN GOLD OSCARS WITH MGM PURCHASE

"There are a lot of the older pictures in our library, but so what? Particularly with colorization."

COLORIZATION TURNED OLD CLASSICS INTO GOLD

"I would maintain that the older pictures have basically stood the test of time forever. We're grossing almost $1 million with (colorized) videocassettes right now.

"We supplement the film library with seasoning, with 'National Geographic,' with Cousteau, with Audubon, with the National Basketball Association package. But at least we have our bedrock, underlying, paid for – exclusive – programming.

"I believe we have the finest single mass of programming anywhere. We've got 3,000 of the finest motion pictures. Of the classic motion pictures, we probably have 35% in the two libraries we own. We have about 35% of all the Academy Awards that were ever given. If you count them, we have 300 Academy Awards."

"No need in sitting – alone on the shelf. Come to the cabaret...da, da, da, da, boom, boom, boom. I'll have a bounce in my step, a smile on my face. Hey, happy-go-lucky me. There'll be a load of compromising. On the road to my horizon. Like a rhinestone cowboy. Riding out in front of a star-spangled rodeo. Like a rhinestone..."

AT FIFTIETH ANNIVERSARY GALA FOR "GONE WITH THE WIND"

"All I can say is thank God they shot "Gone With the Wind" in color... If I were an alien from another planet, this film would be my textbook, my primer on human behavior. Not what happened in Tianenmen Square or the dropping of the Iron Curtain but 'Gone with the Wind.'" (Film rolls):

"It was a land of cavaliers and cotton fields
Called the Old South
Here in this pretty world
Gallantry took its last bow
Here was the last ever to be seen
Of knights and their ladies fair,
Of master and of slave
Look for it only in books,
For it is no more than a dream remembered...
A civilization - gone with the wind."

WITHIN THREE YEARS MGM FILMS HAVE PAID FOR THEMSELVES

"We're like the Allies after Normandy. We have landed. We are there. The beachhead cannot be obliterated."

EYEWITNESS

Board Member: "I have to believe this has worked out better than Ted ever thought it would. Quite honestly, I think this is a very good structure. I don't think you can have anyone with the kind of visionary skills and impulses of Ted and at the same time have a very careful, studied, thoughtful business operator. So what we've got is the good part of Ted still operating, and where the downside used to be, that's limited. When Ted made a mistake, it was a big mistake, because Ted likes to roll the dice, and Ted believes in his insights.

"Ted's capable of going for something and not worrying about the financial consequences. That's what visionaries do. But if you can combine a visionary with an operating person – well, that's Ted and the board."

NOW ONE OF THE BIG BOYS

"The one thing that we're going to be doing now is to perhaps move a little more slowly and deliberately than we did in the past. That doesn't bother me. In a way, it's kind of nice.

"In the past, when you're small and you're weak, you have to move fast because that's the only way you'll stay ahead of the Big Boys. WE'RE one of the Big Boys now."

RUPERT MURDOCH CHALLENGES CNN WITH AN ALL NEWS NETWORK

"I'm looking forward to squishing Rupert like a bug."

TURNER BECOMES LARGEST STOCKHOLDER AND VICE CHAIRMAN OF TIME WARNER

Press: "For Ted Turner, the merger (of TBS and Time-Warner) liberates his lofty ambition from the shackles of cash-strapped circumstances. After years spent as a relative small-timer, the mercurial entrepreneur finds himself vice chairman of Time Warner, at the center of the world's largest programming engine."

CHAPTER TWELVE
TEN PLEDGES TO THE PLANET

ONE

"I think our major responsibility is to our communities and to our species and to other living things on our planet and to the world at large.

TWO

"I love and respect Planet Earth and all living things thereon, especially my fellow species, mankind.

THREE

"I promise to treat all persons everywhere with dignity, respect, and friendliness.

FOUR

"I promise I will add no more than two children to the Earth.

FIVE

"I promise to use my best efforts to help save what is left of our natural world in its untouched state, and to restore areas where practical.

SIX

"I pledge to use as little non-renewable resources and as little toxic chemicals and other poisons as possible and to work for their reduction by others.

SEVEN

"I promise to contribute to those less fortunate than myself, to help them become self-sufficient and to enjoy the benefits of a decent life, including clean air and water, adequate food and health care, housing, education, and individual rights.

EIGHT

"I resent the use of force, in particular military force, and back the United Nations arbitration of international disputes.

NINE

"I support the total elimination of all nuclear, chemical, and biological weapons of mass destruction.

TEN

"I support the United Nations and its efforts to collectively improve the conditions of the planet."

CHAPTER THIRTEEN
THE "SAVE THE EARTH" CAMPAIGN

"We're really a pretty terrific species. We're the top drawer. And we can be so kind and loving and helpful to each other. We just have to learn to live together like brothers."

PRESS

Broadcasting Magazine: "Turner has been talking about television as a tool for improving the world ever since he rose to prominence in the television industry in the late 1970's. Turner said his push to become a major force in the television business was not motivated by money or power. 'I'm doing it as a crusade.'"

CRUSADING BY SATELLITE TO SAVE THE EARTH

"I'm not concerned any longer about myself. Or money. Or power. I'm concentrating on the biggest problems the world has. I'm trying to get bigger so I'll have more influence. It's almost like a religious fever."

"Like Charlemagne, I'm saving Christendom from the Infidels."

CRUSADING IN PERSON TO SAVE THE EARTH

"Population growth is analogous to a plague of locusts. What we have on this earth today is a plague of people. Nature did not intend for there to be as many people here as there are. (Medical science) eliminated the diseases that were nature's way of controlling numbers and keeping everyone in balance. And like a bunch of termites, we're just going across the world, cutting down trees, changing the ecology, siphoning coal and oil out of the earth's surface and burning it in massive quantities and polluting the atmosphere and messing up the ozone layer...

"And then, on top of that, now we've gone into nuclear weapons and the capability of fighting our wars; the whole issue of warfare is to improve and increase firepower. Now we've increased firepower to the point where we can blow up the entire world. We ought to be in daily communication with the Russians about disarmament. We ought to be working like beavers on alternate energy.

"We got to the moon; we should learn to live on this beautiful planet we've got here until the sun burns out instead of ending it. We have two alternatives, humanity does: we can either make the world into a Garden of Eden or we can go fight the battle of Armageddon. I'd like to see it be the former."

"The next ten years are going to be very difficult. Probably the most difficult man has ever faced...

"The planet is collapsing – the ecosystem is collapsing under the sheer weight of five billion people."

"It's a matter of survival, number one; and number two, a matter of good business."

THE "SAVE THE EARTH" CAMPAIGN

EYEWITNESSES

Mike Gearon: "The president of the United States does not get the treatment Ted gets in a lot of places..

"There's no one else who is as much a citizen of the world as he is. He's a one-world person, but his idealism has totally to do with his business interests. He's not sacrificing a single idealistic principle for his business. It's totally in synch."

Tom Belford, Director Better World Society: "Ted sees the angles on everything. He's the consummate do-good, do-well guy. Ted cares deeply about issues, but he also understands the happy synergy: if you aspire to be a global communicator, it makes sense to be associated with world peace and the environment. He could raise the profile of his company by connecting an issue agenda to a business agenda. Most journalists bend over backwards to avoid that, but not Ted."

LOSES ADVERTISERS TO TAKE A POSITION, AIRING "ANCIENT FOREST: RAGE OVER TREES" AND "ABORTION FOR SURVIVAL"

"You bet your bippy we're taking a position. We'll just take a lick for a night. So it'll cost us four or five hundred thousand dollars...

"I don't want anybody else telling me what my daughter has got to have, or my wife, or my girlfriend. We live in a free country. There's absolutely no way that's anybody's business but the person that's involved. That's my opinion. We'll give the other bozos a chance to talk back. They look like idiots anyway. That's fine if those people don't ever want to have sex. Swell. I happen to enjoy it. I don't get near as much as I want to."

ON THE ENERGY CRISIS

(1970's) 'This is the year I was going to get rich, pay off all the debts. Now I feel like Napoleon in Russia waiting for the other guys to surrender. Then this soldier comes in and tells me they haven't given up, they're still out there, and besides, it's starting to snow. And now for the bad news; there's no food.''

ON DESTRUCTION OF PLANET'S RESOURCES

"What is a newspaper? You cut down a tree, mash it, make paper, deliver it by truck to somebody's house, then another truck takes it away – and every stage requires fuel!"

"I grew up hating newspapers, because my father used to come home at night with a newspaper tucked under his arm, editorial after editorial about banning billboards."

CNN

"a newspaper you can watch."

ON PEAT BOGS

(1979, lunch with Sam Zelman at Stadium Club): "You know, all you have to do is keep up with what's going on. Anybody with a little money can make a lot of money. Just like the other day I was reading that there's a fuel shortage. Well we know that the world is running out of oil, and that we're having trouble providing all of our energy needs.

"My idea and I'm doing it, is to buy up some peat bogs, because eventually they'll have to burn peat to create energy. So I have my eye on thousands of acres of peat bogs in Florida. And that's all it takes. You just have to be a step ahead of everybody else...

THE "SAVE THE EARTH" CAMPAIGN

"I enjoy life. I find these interesting projects to do – and the world's my oyster!"

MOST OF HOPE PLANTATION'S 5000 ACRES IS ACTUALLY 12 FOOT DEEP PEAT BOGS

"There's peat on that land. There's a peat bog. They thought they were selling off a plantation, but I was buying a peat bog."

UNITED NATIONS EARTH SUMMIT PROMOTION SPEECH IN NEW YORK

"With the increasing numbers of people plus the fossil fuel use, we're putting tremendous pressure on our little home here. The numbers have tripled during my lifetime, and the size of the planet hasn't.

"We have to PLAN for our futures. We HAVE to think globally. It's not just Fortress U.S.A. We can't just pull up the drawbridge. We all breathe the same air...

"You know, we want to make everyone here aware the US is dragging it's heels." (WTBS provided a "Save the Earth" documentary series for the summit and international broadcasting)

TURNER IS KEYNOTE SPEAKER AT UNITED NATIONS EARTH SUMMIT IN RIO DE JANEIRO

Eyewitness, Director of the United Nations Earth Summit, Maurice Strong: "Ted was ten years ahead of everybody else on the environment. Now he's leaving it up to others to follow through."

EYEWITNESS

Ira Miskin: "I think that Turner takes it very personally and feels very deeply criticism about things he tries to do that people think of as 'image parading.' All of his involvement with the Soviet Union, for example, The Better World Society, which many people felt was a joke, was a very, very serious endeavor for him.

"These are all outward manifestations that are very Turneresque, in which the execution of the ideas ran away with, or didn't live up to, the underlying vision of what he wanted to do. He feels, I think, very deeply about the world, and relationships among people...

"It's what binds you to him and that company. You finally believe that what he's telling you to do comes from every molecule of his fabric. And that he will stick to it until his last drop of blood."

"SAVE THE EARTH" IS A SERIOUS PASSION

"I see the whole field of environmentalism and population as nothing more than the survival of the human species."

CHAPTER FOURTEEN
DREAM-O-VISION, TV AT GROUND ZERO

TURNER IN MIDDLE EAST IS STRUCK BY VOLATILE TENSION BETWEEN NEIGHBORS

Kathy Leach (six months before Persian Gulf War): "He started thinking about it. The whole trip he was trying to figure out how to bring peace to the Middle East – how could he bring peace to the Middle East?"

WHITE HOUSE TRIES TO GET ALL PRESS OUT OF BAGHDAD

WHITE HOUSE STAFF (calling CNN) "Your people are in serious danger."

TURNER (from ranch in Montana to CNN staff): "We have a global job to do and we should do it...

"Those who want to come out can come out, but those who want to stay can stay.... They can stay. You will not overturn me on this."

TO BERNARD SHAW IN BAGHDAD

"As far as I'm concerned you're all HEROS!

"You're heros for being there up to now, but especially now since it's January fifteenth where you are. I'm very proud of you."

TO ROBERT WEINER IN BAGHDAD

TURNER: "You're doing a helluva job over there, Robert. Congratulations!"

WEINER: "Thanks, I think it's important that we're here."

TURNER: "We HAVE to be there. We're a global network! Also, if there's a chance for peace... personally, I'm afraid it's too late for that, but if there is, it might come through us. Hell, both sides aren't talking to each other, but they're talking to CNN. We have a major responsibility."

WEINER: "The situation doesn't look too promising. We are prepared for the worst."

TURNER: "Well, keep up the good work, and best of luck."

EYEWITNESS

Robert Weiner, CNN producer: "If we could somehow pull it off it would be the journalistic equivalent of walking on the moon. To cover a war live in real time, from behind enemy lines in the enemy's capital!"

EYEWITNESS

Ed Turner, CNN, Atlanta: "It's as though we were in training all this time for just this story."

EYEWITNESS

Robert Weiner: "CNN's countdown to the deadline was extraordinary television. Virtually all the network's resources were mobilized as CNN hopped around the globe from the United States to the Middle East, to Europe, the Soviet Union and Asia. With live and taped reports, interviews and analysis, official pronouncements and reaction from people on the streets, the network captured the mood and the moment precisely. CNN's reach was impressive but as far as I was concerned there was only one dateline that counted; and whenever an anchor in Atlanta intoned, 'We go live now to Baghdad and CNN's Bernard Shaw,' I felt I was living history."

EYEWITNESS:
JANUARY 16, 1991, 11:05 A.M.

LARRY DOYLE, CBS (running through Al-Rasheed Hotel): "This is it... we're outta here. The New York desk just got the call from Washington... so did ABC. It was a special code: 'Everyone's fine at home except the kids have the sniffles.' That means it's coming down tonight – did your guys get it in Atlanta?"

ROBERT WEINER: "No one in the CNN home office had ever mentioned a code to me."

PETER ARNETT, CNN anchor: "Listen, guys. Absolutely nothing's changed between yesterday and today except for Fitzwater's announcement (for all press to leave Iraq). It's obvious the White House doesn't want us here. They don't want any reporters here, but that doesn't mean we should panic. I say we stay and that's that."

PILOT ON U.S. AIRCRAFT CARRIER WARNS PRESS WHO THINK AL-RASHEED HOTEL IS SAFE

"Are you serious? We're using that hotel as a landmark when we go for the bridges. We're going to come right over it."

WALTER CRONKITE WARNS BERNARD SHAW ON LIVE TV

"The decision to stay in a place that is clearly a major danger zone where one's mortality has to be considered on the line is probably the toughest decision that any newspaperman or reporter ever had to make...

"I suppose there comes a point where it becomes foolhardy to risk one's life to do that job if it's almost certainly fatal at the end."

EYEWITNESS

NAJI, Iraqi contact: "At present, we do not have the means to properly care for many reporters.

"CNN may stay. Your coverage has always proved impartial and balanced. From your first days in Baghdad, you have always been fair."

ROBERT WEINER: "I tell ya, this is going to provoke a shit storm with the others."

NAJI: "That's their problem. We are granting this permission only to you."

MRS. NIHAD, hotel employee: "Oh, please, Mr. Robert, if you leave we will all die... The only reason the hotel has not been bombed is because you are here. With no journalists they will kill us all."

NINETY MINUTES LATER BAGHDAD COMES UNDER ATTACK

BERNARD SHAW (lunging for four-wire phone as sky flashes red): "This is Bernie Shaw. Something is happening outside. Peter Arnett, join me here. Let's describe to our viewers what we're seeing...

"The skies over Baghdad have been illuminated...

"We're seeing bright flashes going on all over the sky....Peter."

PETER ARNETT (on floor under table): "Well, there's anti-aircraft gunfire going into the sky...

"However we haven't yet heard the sounds of bombs landing but there's tremendous lightning in the sky, lightening-like effects... Bernie."

ROBERT WEINER: "God, I hope Atlanta is hearing us." (as the anti-aircraft batteries on the rooftops around the hotel opened up, filling the sky with a barrage of hot white light.)

BERNARD SHAW (now huddled under the table): "I have a sense, Peter, that people are shooting toward the sky and they are not aware or cannot see what they are shooting at. This is extraordinary. The lights are still on. All the streetlights in downtown Baghdad are still on..."

"We're getting starbursts in the black sky.... Peter."

PETER ARNETT: "Now the sirens are sounding for the first time..."

"Turn the lights out!..."

"We are trying to get the lights out in our hotel yet the streetlights are on, and the firing is continuing, and the sirens are continuing..."

"Here with us now is John Holliman."

JOHN HOLLIMAN (who had just run into the room): "Good evening, gentlemen, or rather, good morning. I cannot see any aircraft in the sky here..."

"A lot of tracer bullets going up in the sky, but so far no planes..."

PETER ARNETT (now off the air): "The four wire's down. The four wire's down! Quick! Plug in a battery."

A VOICE IN THE DARK: "Where the hell is a battery?" (They finally got back on the air and continued into the night for seventeen hours straight coverage. Bernard Shaw eventually became so bleary-eyed he trailed off into a stream of consciousness, and had to be switched off the live feed by the Atlanta producers.)

BACK IN ATLANTA

Tim Eason (next morning to "The Boys from Baghdad"): "You wouldn't believe it. The phones have been ringing off the hook. We made history!"

Paul Amos (when TV affiliates in Britain, Germany, France, Israel, Brazil, Sweden - AND the US had dropped their regular programming and picked up CNN instead): "For the first time in history, CNN has beaten all three broadcast networks on a twenty-four hour basis in the CNN universe."

EYEWITNESS

Robert Weiner (After CNN bureau had moved downstairs into the bar): "It began in the distance as a kind of rumbling which grew progressively louder as it made its approach. Within seconds the rumbling turned into a high-pitched whistle that whooshed through the air...

"I had barely taken a step (away from the window) when a deafening explosion, louder than anything I had ever heard before, shattered the plate glass and sent me hurtling across the bar. In that brief instant I was literally picked up and thrown from one side of the room to another. I landed on my stomach with my arms outstretched...

"We raced down to the shelter as if our lives depended on it...

"This was not the first time I had been under fire, nor the first time I could have been killed, but the sheer force of the Tomahawk cruise missile that impacted on the grounds of the Al-Rasheed, blowing out or at least shattering every window in the lobby and bar, brought me face-to-face with my mortality like nothing I had ever gone through before."

WHEN TURNER WAS CRITICIZED FOR ALLOWING SADDAM HUSSEIN TO WATCH CNN

"Of course the Iraqis are the enemy of the United States, but CNN has had, as an internationalist global network, to step a little beyond that....

"We try to present facts not from a US perspective, but a human perspective.... I'm an internationalist first, and a nationalist second."

EYEWITNESS

Robert Wussler: "Ted is an internationalist. Ted believes very strongly in one world, that we are all citizens of planet Earth.

"He wants his company to be as global as possible. He wants to prove that private industry can sometimes do things quicker, better, than governments can."

ON PETER ARNETT, PULITZER PRIZE WINNER

Morley Safer: "Arnett is probably the toughest, fairest, most consistent of all the hundreds of reporters who covered the (Vietnam) War."

EYEWITNESS

ROBERT WEINER (in palace in Amman, Jordan, where the King and Queen had entertained Ted Turner only six months earlier): "Both were intensely curious about the situation in Baghdad and obviously concerned about the long term ramifications of the war."

KING HUSSEIN: "There is a perception in the West that I supported Saddam when he invaded Kuwait. This is simply not true. Jordan condemned the invasion and the violence that followed."

QUEEN NUR: "We are caught in the middle. We share a common border with Iraq and most of our population is Palestinian. There are certain political realities..."

IRAQIS FINALLY GRANT PERMISSION FOR FLYAWAY DISH (PORTABLE EARTH STATION) SO CNN CAN SEND AN UNCENSORED SIGNAL

Robert Weiner: "I want this truck painted like a circus wagon. Candy-colored stripes, stars.. even a big yellow sun. Stencil CNN all over it. When this puppy goes down that road, I want every Allied pilot to know what it is."

EYEWITNESS

Major General Perry Smith, CNN guest military analyst: "By the time the Gulf War broke out, CNN was reaching one hundred and eight countries. By 1990, (a decade after its launch) there were 1700 staffers and seventeen bureaus; by 1991 that had grown to eighteen hundred staffers and twenty-four bureaus. Yet in order to understand how these people pulled the news together it is necessary to examine the technology, the news gathering and sorting process, and the structure and functioning of the main CNN studio in Atlanta.

"One of CNN's biggest strengths is its technical ability to pull together a large amount of information from various places and feed it into the control room in Atlanta. As a result, the producers had a variety of material to choose from, including live and taped shots, live telephone hook-ups and guest experts in Washington and Atlanta who could be on camera in less than two minutes. With a veritable cornucopia of interesting material at their command, these producers could quickly place a vital story on camera.

"In order to turn this juggling act into compelling television, the producers needed a workable but flexible plan. Each day they had to have scripted stories, dozens of guest experts, and a great deal of satellite time with the up-links and down-links to and from the satellites well established. Success required decisive producers who could quickly monitor an enormous amount of information and identify and select the best stories. This diverse menu of choice came about because there were many producers and reporters in the field feeding information to the producers in the control room in Atlanta and telling them about present and upcoming stories.

"A crucial role was also played by highly competent technical people who would say, 'O.K., we've got five minutes satellite time left with Amman' or 'We have ten minutes satellite time coming out of Riyadh at the top of the hour.'

"Producers needed to grasp not only where the best stories were located, but whether they could bring them to Atlanta when the satellite was available. They also had to understand which alternate routes were available if the satellite up-links and down-links were not working properly...

"It seems to me that producers, anchors, reporters, guests, and military analysts need to think more seriously about the contributions they can make to the development and implementation of enlightened public policy. Criticism is a useful journalistic tool, but predictions and suggestions may be even more helpful. Criticism often causes top decision makers to become very defensive, while helpful suggestions are sometimes more readily received and acted upon."

SOLDIERS WATCHED IT, SPIES WATCHED IT, JOURNALISTS WATCHED IT

Secretary of Defense, Richard Cheney: "Our best source of intelligence during the Gulf War was not the CIA nor the NSA, but CNN."

EYEWITNESS

Terry McGuirk, CNN: "Ted Turner's vision, and the guts to stick to that vision, still drive this company."

1991 TIME MAGAZINE MAN OF THE YEAR

Time: "For influencing the dynamic of events and turning viewers in 150 countries into instant witnesses of history, Robert Edward Turner III is Time's Man of the Year for 1991."

CHAPTER FIFTEEN
HOME - AT LAST

TURNER: "I like these ranches. How do you go about getting one?... I think I'll get this one... I'll buy it."

BROKER: "Come on, Mr. Turner, you gotta let me take you there, for gawdsakes. You might not even like it."

WALKING THE WIDE OPEN SPACES AT "BAR NONE"

BROKER: "I gotta warn you. I've lost a couple of clients on this. You're going to see a hell of a lot of rattlesnakes on this place."

TURNER: "Rattlesnakes? Hell, I love rattlesnakes. That's a plus as far as I'm concerned."

AFTER BUYING MORE RANCHES INCLUDING THE "FLYING D"

"Buffalo! I want buffalo as far as the eye can see. Buffalo! I'm going to tear down all these fences and buy a thousand buffalo."

EYEWITNESS

Peter Manigault: "Ted's one of the earliest users, and he is far and away the biggest user in the world, of conservation easements. The biggest easement ever done anywhere was on the 'Flying D,' but every property he's ever gotten, he has put an easement on. The minute he gets a property he tells his in-house lawyer to crank out an easement."

EYEWITNESS

Bob Keisling: "The Flying D is a handsome, high quality place and deserving of Turner's conservation intentions. He has demonstrated in both word and deed that he is a strong conservationist. He will make a good neighbor and a nice addition to the Montana community.

"Bozeman is extremely lucky Ted Turner's the one who is coming in."

TEACHING INDIAN GAME WARDENS AND BOZEMAN CATTLE RANCHERS HOW TO RAISE BUFFALO

"That's what everyone wants. Get a great idea and hit the big time. It's like Edison with the light bulb. Or Henry Ford with the Model T. Everyone wants to come up with the new thing. The great new idea that lets them hit the big one. At least, that's what I've always wanted to do. To try something different than what's been done before. Create a new thing and make a killing.

"I just never liked cattle even though my Dad had them. They trampled down all the grass, and that ruined the cover for quail, which I loved to hunt."

WHY BUFFALO?

"I just like watching them. They were also nearly extinct – and I want to make sure that doesn't happen again. I also intend making twice as much money from bison as you would make from cattle. Unlike cattle, bison don't need hay in winter, except for the buffalo calves, and then only when the bison cow is carrying a second calf. Bison have a higher tolerance for cold weather. Their metabolism rate slows and they require less forage. Their thicker coats better insulate them. They don't ruin rivers and stream banks like cattle. They don't wreck tree stands. They stick to the plains, grazing in the fiercest sun, not sheltering in shade. And they're cleaner – they wipe their buns."

BRUCELLOSIS THREAT TO CATTLE?

"If they get out, naturally I'll be responsible."

EYEWITNESS

David Schwab: (directing two archaeological digs) "This land has unleashed a spiritual power for some of the (Blackfoot) Indians involved with our project. Looking out over the buffalo grazing on the rehabilitated natural grasslands of the 'Flying D,' they feel they are looking at the land as it was before the white man arrived."

"FLYING D" HOSTS 1800 WILD TROPHY ELK

"The only reason those elk are there at all is that they aren't massacred. Without the cattle there will be enough grass left for them to eat."

"FLYING D" IS OFF LIMITS TO HUNTERS

"I bought the place to get away from people. If I wanted to be around people, I would have stayed in Atlanta. The only way to get access is to do like I did and work forty or fifty years and make twenty-two million dollars and then go buy it for yourselves. Me, I want to live as far away as I can from everybody. I'm becoming a hermit. There's nothing wrong with that."

ON RELATIONSHIP WITH JANE FONDA

(On Larry King Live) "As far as I can see, it's going great... Not a very long answer, but you know I'm not used to really talking much about my personal life...

"I've lost eighteen pounds without really going on a diet as such. I'm working out a whole lot more than I did before, and I feel a whole lot better. I've got to say that. Regular exercise really is a good thing – and everybody can work out with Jane with this tape... it's a rare opportunity – it's an opportunity that can be shared by everybody, and it's really a heck of an experience."

"We have a community of interests. She's certainly been working on these (environmental) issues longer than I have, but I've been working very hard on them in the past decade."

ON FLY FISHING

"I find it to be my second most favorite thing to do – but a reasonably close second."

ON MAKING MOVIES

"You've got to start with a story. Great story, great book, great script. It starts with the writing."

MAKES FIVE MOVIES IN MONTANA

"I LOVE movies. I hope to be a major force in the movie business. I love life. I love the planet. I love my wife, my kids. Animals. I love albatrosses, eagles – chipmunks! I love trees. The redwoods in California!"

A BORN SCORPIO

Astrologer: "If Ted Turner didn't have Libra rising to counter his main sign, which is Scorpio, he would be pure awful. But the Libra gives him that childlike, innocent quality. The little-boy aspect makes people want to protect him. They can understand his game, and relate their fantasy to his."

MAKES "GETTYSBURG," LIVING OUT CHILDHOOD FANTASY OF DYING AT PICKETT'S CHARGE

"There wasn't a family in the United States that was untouched by Gettysburg. Both sides, the North and the South. Everybody else turned this film down. EVERYBODY – for fifteen years. But I liked this film. I didn't like it. I LOVED it. I know the Gettysburg Address almost by heart. It's the most significant three days in the history of our country... probably. If it had gone the other way, we'd probably be two countries today. Like Boznia-Herzegovina, we'd still be duking it out.

"When I moved to the South they were still fighting the Civil War – Big Time. I became a Confederate to survive – and for a Confederate – after seeing 'Gone With the Wind' – that made a big impact on me – the only place you wanted to die was Pickett's Charge at Gettysburg, where they thought they were going to win the day..."

"When I was a little kid, about nine or ten, I dreamed of dying in Pickett's charge."

"So I said, 'Well, I've got to have a cameo when we're going to do Pickett's Charge.'"

WATCHING THE BATTLE SCENE IN 'GETTYSBURG' FROM UP HIGH ON A CHERRY PICKER

(He and Jane Fonda were so moved they cried) "I'll tell you how I felt. It's like the way you felt on your wedding night or when your father died."

DOES JANE FONDA HAVE A ROLE IN HIS MOVIE MAKING?

"Absolutely! I'm telling you! Does Jane have a role? She's my Scarlett O'Hara. She's my sweetheart. And she knows a lot about movies. Forty-nine of them she acted in. She produced ten or eleven. One of which was 'On Golden Pond.' One of which was 'China Syndrome.' They were movies that made a difference. I'm inspired by her and her dad, and Humphrey Bogart and Sam Elliot... and Martin Sheen. We watched 'Apocalypse Now' last night. We want to make good movies of all kinds.

"I like comedies. I like adventures. I like epics. I like Westerns. I like all kinds of movies. I like science fiction. I loved 'The Day the Earth Stood Still.' I love 'The Thing'... '2001'... I like it all! I like it all!..."

"I don't read scripts very much. I read them occasionally. I let Jane read them, and she can tell me 'Good, OK, needs more work.'

"She's producing a movie right now called 'Lacota Woman,' which is about the more recent battle at Wounded Knee. We're doing this series of television movies on the American Indian. She has worked on that script with the writer and director for six months. EVERY day."

TURNER PICTURES, CASTLE ROCK, AND NEW LINE CINEMA ARE CHURNING OUT MOVIES FOR TV AND THE BIG SCREEN

"We're going to do 'Joan of Arc' next. It's been done before, but we're going to use a fifteen year old girl, which she really was at that time."

"I had MGM in 1985 but only for two months. I never got to green-light a movie or anything. I had to sell it. I knew I had to sell it because I was going to go broke if I didn't... I didn't want to go out. I wanted to make movies...

"And now we're into theatrical movies. I want to make as many great films as possible."

"Show biz! There's no business like show business. No business I know. Everything about it is appealing. Da-da da-da da-da da-da daaa!"

"FLYING D" IS HOME

"I am spending more time in Montana than I should, and a lot less than I'd like to.

"I've invested a lot of bucks turning this land back to the way it used to be. More than I thought I would, but it's worth it."

EYEWITNESS

Teddy Turner (oldest son): "The Flying D is really becoming the new family home. We get together for holidays – but it's a big crowd. At Thanksgiving last year it was over twenty, all family. Jane and my father have really made a push to get the family – not BACK together, but together. At Christmas, Thanksgiving, spring breaks.

"And, of course, the Braves have been doing pretty well lately, so we all get to see each other at the games."

EYEWITNESS

Mike Gearon: "I think Ted is finally convinced he's rich. He used to say to me, 'What do you think about how I'm doing today? How would you measure it in, say, fifties economic terms? I think that what he was really saying was, 'How do I match up to what my dad did?' But now, he's finally convinced – it's not all going to go down the drain."

"My father had this burning desire to succeed. Success, success, success! And I went to this military school where they pounded in, 'You got to get to the top, boy, you got to get to the top.' Well, I got to the top."

FONDEST WISH NOW THAT TURNER IS A MULTI-BILLIONAIRE?

"If I had only one wish for something that would give me personal joy and satisfaction, it would be to have my father come back and show him around. I'd like to show him the whole shooting match... I really would. I think he'd really enjoy it."

TURNER FAMILY FOUNDATION

"I want to get my children in the habit of giving money away instead of spending it."

FORBES MAGAZINE

"Turner Foundation's $150 million endowment is to be expanded to $500 million within next year (1997): Green Causes, please apply." (In 1996, Foundation gave money and guidance to hundreds of environmental causes including Worldwatch, the Bat Conservation Society, and Friends of the Wild Swan.)

TURNER FAMILY FOUNDATION FOUNDS HEMP MUSEUM IN KENTUCKY

"Old Ironsides (the U.S. frigate "Constitution") had sails and shrouds made of hemp. Thomas Jefferson grew hemp on his farm. He and George Washington recommended that the early colonists grow 'the abundant weed' and cultivate it into such products as lamp oil, flour and fabrics for uniforms and clothing. In Virginia growing hemp was mandatory. Betsy Ross made the first American flag, 'Old Glory,' from hemp.''

PRESS ON LEGALIZATION OF HEMP

"Its use predates Christopher Columbus, but this easy to grow plant fiber is capable of replacing wood as the raw material in paper, grows without the use of pesticides or herbicides and is one of the most versatile resources of our time."

TURNER FOUNDATION DIRECTOR

Peter Bahouth, former director of Greenpeace: "It's appalling to me that we don't use this crop. Absolutely appalling. We need to build a constituency of people who understand the benefits of hemp and other alternatives to deforestation...

"Misinformation from the Drug Enforcement Administration is what keeps the United States from using this crop to its greatest industrial advantage. The Turner Family Foundation is very open to promoting the idea that this is a beneficial crop, with no drug related problems, that can even help our nation's farmers out a bit...

"I have no doubt in my mind that hemp will be grown legally in the United States within two years, and because there is such a huge demand for domestic hemp-based products, whichever state first decides to grow it will reap huge benefits...

"One of the things I like about what's going on in Kentucky is that it involves a lot of very conservative people, older people who have credibility on these issues. I mean, legislators support it, there are state resolutions being written for it, there's a lot of political support for hemp in Kentucky. It is absurd that we are disallowed by the government to grow hemp, while the taxpayers are subsidizing the logging of our national forests."

"Reducing our demand and use of wood can save forests, create more jobs and reduce pollution. The U.S. uses more wood per capita than any other country, each citizen using 50% more paper products (packaging, etc.) than those citizens of the next largest consumer, Japan...

"Besides wood demand for fuel and buildings, paper consumption continues to outstrip forest capacities. Last year, 50 million trees were used for pulp, paper and paperboard production in the U.S. And the global demand for wood fiber to make paper increases every year. The Food and Agriculture Organization predicts the doubling of world demand for paper within 15 years. Unfortunately, about 60 percent of the paper Americans use for stationary, phone books, newspaper, and the like, ends up in land fills. Society is systematically turning resources into waste."

US NEWS AND WORLD REPORT

"You can't get high smoking jeans made from it. The oil tastes pretty good on salads, but don't expect to get a buzz from it... Some estimates put worldwide trade in hemp products at $100 million last year, a figure that experts say could double or even triple in the next few years."

Marilyn Craig, Business Alliance for Commerce in Hemp: "Hemp is literally capable of saving the planet."

'ROUND THE WORLD ON THE WEB

"September 19, 1997 Web posted at: 12:10 a.m. EST (0510 GMT) NEW YORK (CNN) — CNN founder and Time Warner vice chairman Ted Turner announced Thursday night that he will donate $1 billion over the next decade to United Nations programs.

"Turner made the announcement at a dinner held in New York by the United Nations Association-USA to honor Turner for his contribution to the international community. He was presented the Global Leadership award by the group."

"What I'm trying to do is set a standard of gallantry. The world is awash in money, with peace descending all over the Earth. We can make a difference in the future direction of the planet."

"This is not going to go for administration. This is only going to go for programs; programs like refugees, cleaning up land mines, peacekeeping, UNICEF for the children, for diseases, and we're going to have a committee that will work with a committee of the U.N. The money can only go to U.N. causes... The donation will be made in 10 annual installments of $100 million in Time Warner stock."

WHY ONE BILLION?

"A billion's a good round number... I made the decision to donate the money only two nights ago. It was based on the increase in my net worth since the beginning of the year.

"When I got my statement in January, I was worth $2.2 billion. Then I got another statement in August that said I was worth $3.2 billion. So I figure it's only nine months' earnings – who cares?"

ON LARRY KING LIVE

TURNER: "I'm no poorer than I was nine months ago, and the world is much better off... The programs will focus on jobs, land mines, education and global warming."

KING: "Are you sure we have global warming?"

TURNER: "Haven't you been outside lately? It's hotter than hell out there...

"The more good I do, the more money has come in. You have to learn to give. You're not born to give. You're born selfish.

"The decision was to give the money based on the value of the stock, but that it is possible the stock will increase in value. Who knows, I might get a little money back from this deal."

KING: "Did you talk to Jane Fonda about this?"

TURNER: "Yeah, two nights ago when I thought of it in a hotel room here in New York. (momentarily silent) It brought tears to her eyes. She said, 'I'm proud to be married to you.'...

"I plan to be a fund raiser for United Nations causes – so everybody who is rich can expect a call."

CREATES A PHILANTHROPIC COMPETITION

"I'm putting every rich person in the world on notice that they're going to be hearing from me about giving more money away..."

"There are so many rich guys in the world, billionaires. The world is awash in money and nobody knows what to do with it. We don't want the money they KNOW what to do with, just the money they DON'T know what to do with...

"What good is wealth sitting in the bank? It's a pretty pathetic thing to do with your money."

PRESS

"Media mogul Ted Turner suggested last year if charitable giving were made more competitive the rich would eagerly open their purse strings. In addition to the Forbes 100 List of the wealthiest

HOME - AT LAST

Americans, he proposed an 'Ebeneezer Scrooge Prize' for the biggest tight-wads and the 'Heart of Gold Award' to honor the most generous. Turner's donation was the single largest ever made by one individual."

TURNER AS JIMINY CRICKET FOR AMERICA

(Leaning chin on his hands, looking dreamily at a porcelain figure of Jiminy Cricket and Pinocchio): "It's one of my favorite things. Yeah, I want to be like Jiminy Cricket for America, its conscience. Remember how Jiminy Cricket always told Pinocchio to go to school, and to do the wise things?"

THE MOST MULTIFACETED MAN I EVER MET

Press: "Turner says that a magazine columnist once called him the most multifaceted individual he had ever met. Multifaceted!

"And this guy had been all over. He had interviewed athletes, lawyers, doctors, musicians, politicians. He said Turner was this magazine's top guy. Their TOP guy. Multifaceted. Damn!"

FROM CHILDHOOD TURNER WANTED TO BE EITHER AN EXPLORER OR A HERO. NOW?

"I want to be the hero of my country. I want to get it back to the principles that made it good. Television has led us in the last twenty-five years down the path of destruction. I intend to turn it around before it's too late.

"I'll tell you what I want to do. I want to set an all time greatest achievement record, greater than Alexander Graham Bell or Thomas Edison, Napoleon, or Alexander the Great. And I'm in a great position to do it, too."

"I want to be remembered as somebody who made a difference."

CONCLUSION
PRINCE OF THE GLOBAL VILLAGE

When Turner was named 1991 Man of the Year, Time Magazine's cover story, "Prince of the Global Village," paid tribute to him with these words:

"Visionaries are possessed creatures, men and women in the thrall of a belief so powerful that they ignore all else – even reason – to ensure that reality catches up with their dreams.

"The vision may be the glory-driven daring of a Saddam Hussein, who foolishly tried to extend his rule by conquest and plunder, or the seize-the-day bravery of a Boris Yeltsin, who struggled to free a society from seven decades of iron ideology. But always behind the action is an idea, a passionate sense of what is eternal in human nature and also of what is coming but as yet unseen, just over the horizon...

"A generation ago, social theorist Marshall McLuhan proclaimed the advent of a "global village," a sort of borderless world in which communications media would transcend the boundaries of nations... For influencing the dynamic of events and turning viewers in 150 countries into instant witnesses of history, Robert Edward Turner III is Time's Man of the Year for 1991."

RIDING A WHITE HORSE

Ted sees through his own kaleidoscope. In the simple act of looking out his office window, Ted can see hills that still have shell marks from the Yankees' cannons, his friends; Jimmy Carter and Mikhail Gorbachev, and the rest of the world.

On "Larry King Live," Jane Fonda said what surprised her most on meeting Ted Turner was his humor. He colorizes everything he touches – often in gold.

Ted sees color where others see black and white. He sees a round planet where others see flat countries with borders. He sees energy-giving peat bogs where others see plantations. He sees free ranging buffalo where others see environmentally destructive cattle. Where others see the enemy, he sees guys that are just lookin' to be our friends.

Ted's horizons stretch like the rubber band in a little boy's slingshot who believes his stone can bounce to space and back and make a difference. He fears no giant. Ted's shots are aimed to have a ripple effect that will continue long after he is gone. Each of his creations operates without him, so they can keep going until CNN plays "Nearer my God to Thee."

As he grew, his missions grew. First to save hurt and abandoned creatures – to save fish, then birds with broken wings, stray dogs, then habitats for endangered species, then rain forests... the ozone... the oceans, timberlands... marshlands... and his fellow species, mankind... and even their habitat, the planet. He had to save his species from self-destructing by promoting peace on their planet by the year 1 A.P.

CONCLUSION

When Ted Turner was ten years old, another poem by Oliver Wendell Holmes he loved and memorized was "The Chambered Nautilus." This is the last stanza:

> "Build thee more stately mansions, O my soul,
>> As the swift seasons roll!
>> Leave thy low-vaulted past!
>
> Let each new temple, nobler than the last,
> Shut thee from Heaven with a dome more vast,
>> Til thou at length art free,
>> Leaving thine outgrown shell
>> By life's unresting sea!"

Just as the chambered nautilus grew, each of Ted Turner's missions became "nobler than the last."

He was given the title of "Terrible Ted," "The Mouth of the South," "Turnover Teddy," "the Capsize Kid," "Captain Courageous," "Captain Outrageous," "Captain Comeback," "Maverick with a Mission," "Wide-eyed Do-good-er," "Crusader by Satellite," and, by his sailing buddies, "The Voice of America." But the only title he ever wanted was either explorer or hero.

Jacques Cousteau earned the title explorer. There's just one title left for Ted – hero. Hero of the information age who used his power well.

Ted earned this title the old fashioned way – by working hard for it – by setting his course as a child and steering toward it with unwavering speed and courage; just as he sailed a Penguin, always on the edge of capsizing – risking everything for the sake of the goal, always smelling the wind so he knew it was coming before anyone else could feel it.

He says, "I want to be the hero of my country. I want to get it back to the principles that made it good." Yes, hero will do. Swashbuckling hero. AWWRIIGHT!

"The more good I do, the more money has come in."

Ted Turner

AFTERWORD

TED TURNER'S LITTLE INSTRUCTION BOOK
CAPITALISM
WHAT A GREAT SYSTEM – HOW TO WIN

If Ted Turner were to write a self-help book with ten easy aphorisms it might go something like this:

1) READ VOCIFEROUSLY.

2) IDENTIFY WITH YOUR FAVORITE EXPLORERS AND HEROS TO SEE WHAT YOU COULD ACCOMPLISH IF YOU REALLY TRIED.

3) NEVER LOOK BACK. DON'T WORRY ABOUT IT. JUST KEEP MOVING FORWARD.

4) TACKS ARE LIKE SNOWFLAKES: THEY ALL LOOK THE SAME BUT EACH ONE IS DIFFERENT.

5) YOU CAN SMELL WHERE THE WIND IS.

6) ALWAYS THINK SEVERAL CHESS MOVES AHEAD: YOU'LL BEAT A ONE-MOVE PLAYER EVERY TIME.

7) THE RABBIT CAN OUTRUN THE FOX, BUT HE BETTER GET ON HIS HIND LEGS AND HOP.

8) WHEN BRER RABBIT GOT CAUGHT IN THE BRIAR PATCH, HE WAS HOPING HE'D GET THROWN INTO THAT BRIAR PATCH

9) SNEAK ATTACKS, PRE-EMPTIVE FIRST STRIKES. THAT'S THE ONLY WAY THE LITTLE GUY CAN BEAT THE BIG GUY.

10) GET A SWORD.

APPENDIX
1998 NEW YORK GOODWILL GAMES

LOCATIONS, EVENTS, AND DATES
For more information call CNN, 404-827-3400

BATTERY PARK CITY
Opening Ceremonies, July 18

MITCHELL ATHLETIC COMPLEX
Uniondale, Long Island

TRACK AND FIELD, July 19-22
Men's and Women's Events
Decathlon and Heptathlon

MADISON SQUARE GARDEN
New York City

BASKETBALL, July 19 - 24
Men's Team Competition

U.S.S. INTREPID AIRCRAFT CARRIER
New York City

BOXING, July 27 - August 1

WRESTLING, July 25 - 26
Team and Individual Competition

WEIN STADIUM
Columbia University, New York City

CYCLING, July 25 - 26
Men's and Women's Team Competition

GOODWILL GAMES AQUATIC CENTER
Eisenhour Park, East Meadow, Long Island

SWIMMING, July 28 - August 2
Men's and Women's Events

SYNCHRONIZED SWIMMING, July 19 - 20
Duet and Team Events

DIVING, July 23 - 27
Men's and Women's Events

WATER POLO, July 20 - 22
Men's Team Competition

NASSAU VETERANS MEMORIAL COLISEUM
Uniondale, Long Island

FIGURE SKATING, July 29 - August 2

GYMNASTICS, July 19 - 26
Men's, Women's and Mixed Events
Team Competition and Exhibition

RHYTHMIC GYMNASTICS, July 23-24

APPENDIX

NEW YORK HARBOR - BATTERY PARK TO CENTRAL PARK
New York City

TRIATHLON, July 5

Swim: 1.5 km, New York Harbor at Battery Park
Bike: 40 km, Battery Park to Central Park
Run: 10 km, Central Park

CENTRAL PARK
New York City

BEACH VOLLEYBALL, July 22 - August 2
Men's and Women's Doubles

NOTES ON SOURCES FROM RESEARCH

AUTHOR'S NOTE

Page   1   "All these missiles" Sports Illustrated, June 23, 1986
            3   "I wanted to tie the world together" Wall Street Journal, February 1, 1990
            5   "I'd like to be Charlemagne." Christian Williams, Lead, Follow or Get Out of the Way, The Story of Ted Turner. New York: Time Books, 1981 p. 84

INTRODUCTION

7   "The networks are run by a greedy bunch" Lead, Follow p. 17
7   "I just want to welcome you" Hank Whittemore, CNN, The Inside Story, How a Band of Mavericks Changed the Face of Television News. New York: Little, Brown and Co., 1990 p. 124
7   "Now, stick your finger" Sports Illustrated, June 23, 1986
8   "I'm doing it as a crusade primarily." Broadcasting, November 25, 1985
8   "What I'm trying to do is set a standard of gallantry" Wilmington Morning Star, Wilmington, NC, September 20, 1997
8   "At CNN we are flying the flag of the United Nations" Atlanta Constitution, June 2, 1980

CHAPTER ONE
PUNCTURING THE IRON CURTAIN WITH THE GOODWILL GAMES

9	"Goddurn, Wussler. Why can't" New York Times, December 6, 1985
10	"One official said" Robert and Gerald Jay Goldberg, Citizen Turner, The Wild Rise of an American Tycoon. New York: Harcourt Brace and Co., 1995 p. 322
10	"In ten short months" Ken Bastian, Moscow '86 Goodwill Games. Atlanta: The Publishing Group, 1986 p. 19
11	"You know, we're going to" Citizen p. 369
11	"I'm so happy with the way" Wall Street Journal, July 8, 1986
11	"We can best achieve peace" Ibid July 3, 1986
12	"It's the biggest joint effort" Sports Illustrated, June 23, 1986
12	"How much time we got, huh?" New York Times, July 13, 1986
12	"It's easy to depersonalize" Wall Street Journal, July 8, 1986
13	"You tell me who's got the power." Newsweek, February 9, 1987
13	"That's what's wrong with America today." Citizen p. 370
13	"I see myself as a citizen" Ibid p. 368
13	"These games have proven" Moscow '86 p. 182
13	"I don't think I've ever seen Ted so happy" Citizen p. 370
14	"He poured his heart and soul" Ibid p. 425
14	"The 1986 Goodwill Games with the Soviets" Ibid p. 17
14	"I didn't do it to make money." Inside, p. 298
15	"To some he looks too idealistic" Sports Illustrated, June 23, 1986
15	"The Americans will be missed." Moscow '86, inner flap.
15	"I truly believe the Goodwill" Wall Street Journal, July 8, 1986
15	"It was a major mark in history." Ibid
15	"Don't forget that we're doing a story" Citizen p. 371
16	"Turner Broadcasting is the outfit" National Review, September, 1988
16	"Wussler! I asked Georgi Arbatov" Sports Illustrated, June 23, 1976
17	"I love this view." New Yorker, September 12, 1988

NOTES ON SOURCES FROM RESEARCH

18	"Last year I spoke to Moscow University" Inside p. 307
19	"Might does not make right." Sports Illustrated, June 23, 1976
20	"I think it was a success" Broadcasting, August 17, 1987
21	"It's a new challenge" Ibid
21	"I believe that in order to solve" Ibid
21	"The world is a lot less dangerous" New York Times, July 22, 1990
22	"Ted Turner moved briskly" Atlanta Journal-Constitution, July 27, 1994
22	"Satellite." Ibid
23	"We've had some complaints" Ibid
23	"We welcome this opportunity" Ibid

CHAPTER TWO
A BORN CRUSADER

25	"Ay, tear her tattered" Oliver Wendell Holmes, The Best Loved Poems of the American People. Garden City, NY: Doubleday and Co., 1936
26	"I was a vociferous" Paul White Award Acceptance Speech, September 20, 1989
26	"In the summers, because my grades were not good" Roger Vaughan, Ted Turner, The Man Behind the Mouth. Boston: Sail Books, Inc., 1978 p. 153
26	"My father had an idea of what I should do" Citizen p. 27
26	"My father put the screws to me early" Sports Illustrated, August 21, 1978
26	"I started with the fables" Lead, Follow p. 26
27	"I wanted to be a kind of knight in shining armor" Citizen p. 52
27	"When I was a kid I was really upset" Wall Street Journal, June 1, 1980
28	"I got beaten up" Lead, Follow p. 21
28	"It was pretty rough." Tom Snyder, Tomorrow Show, October 24, 1977
28	"I was certainly the only boarding student" Turner address at McCallie, January 14, 1993
28	"At first I was just" Lead, Follow p. 25

29	"He wasn't disliked. He just went on his own little beacon." Porter Bibb, It Ain't As Easy as it Looks, Ted Turner's Amazing Story. New York: Crown Publishers, Inc., 1993 p. 19
29	"I don't think I'd like to go" Citizen p. 39
29	"Well, that's a pretty" Lead, Follow p. 40
30	"In those days I became very religious." Citizen p. 90
30	"I wanted to go to some little Timbuktu" Ibid p. 318
30	"I had worked hard at being the worst cadet." Lead, Follow p. 25
30	"For several years I was" The Man Behind p. 156
30	"I've always felt that McCallie was kind of like my second home." Citizen p. 43
30	"I loved this school a lot" New York Times, February 8, 1994
31	"She was sweet as a little button" Time, August 9, 1982
31	"She came out of that coma" Sports Illustrated, August 21, 1978
32	"For Lack of Water" McCallie School "Tornado" April 17, 1956
32	"I'm Ted Turner from Savannah" Citizen p. 56
32	"There was a lot of bull" Lead, Follow p. 32
32	"Brown was too much like prep school." Newsweek, June 16, 1980
32	"Basically, we both liked to get drunk and chase" Lead, Follow p. 33
33	"I was really happier" Ibid
33	"My dear son" Ibid p. 29
36	"When I got into economics" Brown Alumni Monthly, September 1975
37	"At Brown I was a rebel ahead of my time." East Bay Window, Phoenix, Arizona, September 14, 1977

CHAPTER THREE
LIGHTNING STROKES, SNEAK ATTACKS

39	"I worked fifteen hours" Citizen p. 86
39	"Early to bed, and early to rise" Maria Shriver, First Person, May 6, 1992
39	"He thought billboards were great" Citizen p. 86
39	"Ted traveled with me for about six months" It Ain't p. 34

NOTES ON SOURCES FROM RESEARCH

40	"I think his father was so proud of Ted then." Citizen p. 89
40	"Driving in to work he told me" Time, August 9, 1982
40	"You know, you have the opportunity" Citizen p. 16
40	"People think I'm a crazy man" Sports Illustrated, September 23, 1986
41	"Looking back, for about six months he had" Citizen p. 13
41	"I thought he was having" Ibid
41	"My father could be" Lead, Follow p. 20
42	"Dad just got anxious" Sports Illustrated, August 21, 1978
42	"He put a bullet through his head" Ibid
42	"It was devastating" Citizen p. 17
42	"My father was half way to the big time" Sports Illustrated, August 21, 1978
42	"I want to run the business" Lead, Follow p. 40
43	"He was scared" Lead, Follow p. 41
43	"I was sad, pissed and determined" Lead, Follow p. 41
44	"Lightning Strokes" Ibid p. 12
45	"I also threatened to build" Time, August 9, 1982
45	"The big cheeses out there" Lead, Follow p. 45
45	"By the way, Irwin" Ibid p. 40
45	"I wound up convincing" Ibid p. 45
46	"I want to thank you" Outdoor Advertising Association News, May 1963 p. 5
46	"He was going to hold it all together" Citizen p. 99
47	"Right off, Ted created a sense of paranoia" It Ain't p. 51
47	"Stick with me. If I make it" Citizen p. 116
47	"If things get really bad. I can always kill myself." Time, August 9, 1982
47	"I'd like to point out" Outdoor Advertising Association News, May, 1963 p. 25
48	"What you do is you get a bank" Citizen p. 117
48	"The dream was just to build on the dream" Citizen p. 116
48	"Oh, boy! Was Ted mad." Lead, Follow p. 56
48	"Hell, after about four years" Ibid p. 53
49	"Despite all his other remarkable achievements" Ibid p. 44

CHAPTER FOUR
RADIO AND TV - TALKING BILLBOARDS

51	"My own babies" Atlanta Constitution, January 13, 1976
52	"He's put me in a role" The Man Behind p. 192
53	"There was a teenager with a nasal condition" Lead, Follow p. 55

53 "See, I could take" Ibid
53 "When I bought Channel 17" Inside p. 13
53 "Gawd almighty!" Esquire, October 10, 1978
53 "The guy's really got balls." Inside p. 13
54 "I had never watched the station" Lead, Follow p. 25
54 "Mainly hippies" Television/Radio Age, June 24, 1974
54 "To me, the whole damn thing" Ibid
54 "What makes them so smart?" The Man Behind p. 53
54 "I felt the people of Atlanta" Television/Radio Age, June 24, 1974
54 "The joke was we were number seven" The Man Behind p. 54
55 "Never Look Back." Television/Radio Age, June 24, 1974
55 "Are you a dreamer?" Time, January 6, 1992
55 "In our first meeting, he was amazing." Inside p. 17
56 "You've got to come" Citizen p. 128
56 "It was a complete disaster. The first thing I learned" Lead, Follow p. 52
56 "There were maybe forty-three people" Inside p. 15
57 "No, no, I don't want to do that" Citizen p. 128
58 "The first year I owned" Lead, Follow p. 66
58 "Irwin and the others thought" Ibid p. 67
58 "When I got to Charlotte to head up sales" Inside p. 16
59 "I had to actually go on the air" Esquire, October 10, 1978
59 "I programed the whole station myself in those days." Lead, Follow p. 88
59 "Turner seldom negotiates" Sports Illustrated, August 21, 1978
60 "It's great in the Spring" Roger Vaughan, Grand Gesture, Ted Turner, Mariner, and the America's Cup. New York: Little, Brown & Co., 1980 p. 120
60 "NEWS? What do I want news for?" Inside p. 18
60 "We didn't do the news seriously" Playboy, August 1978

CHAPTER FIVE
THE BRAVES, THE HAWKS, THE FLAMES, THE CHIEFS

61 "I had this horrible" Lead, Follow p. 94
62 "It turned out" Ibid p. 95
62 "A genius? Me?" Ibid p. 111

NOTES ON SOURCES FROM RESEARCH

62	"What do you need to know" Sports Illustrated, June 23, 1986
62	"We televise all the Braves" Grand Gesture p. 96
62	"This is something we HAVE to do" Citizen p. 176
63	"We're going to operate freely" Robert Ashley Fields, Take Me Out to the Crowd, Ted Turner and the Atlanta Braves. Huntsville, AL: Strode, 1977 p. 29
64	"I'm sick of mottos" Ibid p. 39
65	"Make baseball fun again." Ibid p. 228
65	"I never could understand" The Man Behind p. 47 (from Boston Globe)
65	"I bought the Braves because" Atlanta Constitution, April 5, 1976
65	"Awwriight." (moans) "One and nine" Newsweek, June 16, 1980
66	"It takes the edge off life." Ibid
66	"Chief, we gotta grow" Ibid
66	"Pleasure having you here." Ibid
66	"Darleeene..." Ibid
66	"David, you are going to end up the first black" Ibid
66	"Pack up your troubles in your old kit bag" Ibid
66	"This is a new policy for the time being." Take Me Out p. 102
67	"All the coaches were chewing" Playboy, August 1978
67	"What the Hell. I love these guys." Take Me Out p. 114
67	"Get ready to leave tomorrow morning." Inside p. 25
67	"Why did you run to second base?" Ibid
67	"It's a lot tougher out there than I really thought." Take Me Out p. 139
67	"He had never done anything with a ball" Inside p. 25
68	"I left that baseball field" 60 Minutes; Lead, Follow p. 143
68	"That first year I was really active." Ibid p. 141
68	"The first thing I did was spend a million" Playboy, August 1978
68	"Ted has worked very hard since I've been here" Take Me Out p. 220
69	"I'll do it because a lot" Playboy, August 1978
69	"Life itself is a game." Time, August 9, 1982
69	"I'd rather sink than lose." Take Me Out p. 22
69	"If it works it must be right." Grand Gesture p. 77
70	"The fans want to see baseball" Take Me Out p. 44

70	"Not too shabby" Ibid
70	"Oh God, why do I do this?" Ibid p. 153
70	"I beat Tug" Playboy, August 1978
70	"When you're little" The Man Behind p. 227
70	"I never even knew her name." Take Me Out p. 162 (From Atlanta Journal)
70	"All Andy wanted was a no-trade clause" Ibid p. 48
71	"He'll never be traded." Ibid p. 47
71	"Anything you offer Gary Matthews" Lead, Follow p. 139
71	"Everybody is going to be there" Atlanta Constitution, October 22, 1976
71	"There's no law against" Ibid October 17, 1976
71	"Go shake hands" Ibid October 25, 1976
71	"If it were me, I'd be moved" Ibid
71	"This is nice" Ibid
72	"Ted could talk you into anything." Take Me Out p. 227
72	"When they smile blood" Sports Illustrated, August 21, 1978
72	"It's good for my humility" Newsweek, June 16, 1980
72	"Why can't you be like everybody else?" Time, August 9, 1982
72	"A baseball owner's got less rights" Lead, Follow p. 139
72	"If you want to get to the top" Citizen p. 198
73	"Great White Father" Lead, Follow p. 145
73	"Well, there are a couple of things" Citizen p. 192
73	"Just say that for once in his life" Atlanta Constitution, December 10, 1976
74	"The world has gotten along without" New York Times, January 19, 1977
74	"After this is over, you keep that up and you'll get a knuckle sandwich." Newsweek, June 16, 1980
74	"The man questioned my honor" Sports Illustrated, August 21, 1978
74	"Every man suspects himself" Ibid
74	"I started off saying" Lead, Follow p. 141
74	"In a closely contested game" Sporting News, November 10, 1991
75	"Before I got into baseball" Tomorrow Show, October 24 1977
75	"Some things don't need changing." Take Me Out p. 228
75	"I own the worst baseball team" New York Times, January 4, 1977

NOTES ON SOURCES FROM RESEARCH

75	"Outnumbered five to one" Sports Illustrated, August 21, 1978
75	"Ted Turner is every kid" Ibid
76	"Yeah, but I wanted soccer" Lead, Follow p. 124
76	"The Braves, the Hawks" Ibid p. 146
76	"By the time the leagues found out" Ibid p. 124

CHAPTER SIX
CAPTAIN COURAGEOUS – UH – OUTRAGEOUS

77	"I didn't have the ability to play baseball." Bob Bavier, The America's Cup, An Inside View. New York, Dodd, Mead and Co., 1986 p. 77
77	"Ever since he was a little boy" Citizen p. 30
77	"I was so impressed just to be there." The Man Behind p. 77
78	"You can see it. Look out there in the atmosphere." Citizen p. 60
78	"Some races are so beautiful" Sports Illustrated, August 21, 1978
78	"We were on a sailboat about thirty feet long." Citizen p. 77
79	"Ah! The good old time" Sports Illustrated, August 21, 1978
79	"Now look at this" Ibid
79	"If the South had won the Civil War" Time, August 9, 1982
79	"Look, you want to know the faults" Citizen p. 145
80	"I remember him telling me about the New York Yacht Club" Sports Illustrated, August 21, 1978
80	"You're not sailing your own boat" Ted Turner and Gary Jobson, The Racing Edge. New York: Simon and Schuster, 1979 p. 28
80	"You can always change skippers" Brown Alumni Monthly, November 1974
80	"I'm like the grass." Grand Gesture p. 154
81	"You know, when you were fifteen" Playboy, August 1978
81	"He'd get us all together" East Bay Window, September 14, 1977
81	"I mean, there was no way" Lead, Follow p. 150
82	"I can make eleven guys" Playboy, August 1978
82	"We came to win and that's what" Time, September 19, 1977
82	"If I have to watch them lose" Atlanta Constitution, July 6, 1977

82 "Tacks are like snow flakes" Ibid
82 "I really think that man" The Man Behind p. 96
82 "The most fun that you ever have" Playboy, August 1978
82 "We've got 'em!" The Best Defense, Part IX of ESPN series "The America's Cup"
83 "I'm going to do something" Playboy, August 1978
83 "We need to win" The Man Behind p. 77
83 "We didn't use tank tests" Playboy, August, 1983
83 "If being against stuffiness" Time, September 19, 1977
83 "The guy was acting like" Sports Illustrated, August 17, 1977
84 "During the Cup eliminations" East Bay Window, September 14, 1977
84 "I just can't stand snobbery" Newsweek, June 16, 1980
84 "If he's ever selected" Providence Journal, August 20, 1977
84 "There will never be a time" Newport News, September 7, 1977
85 "Gentlemen, congratulations." Ibid August 20, 1977
85 "I've been a very close reader of the sports page" Presidential Papers, Carter Library
85 "I'm happy to be alive" Playboy, August, 1977
85 "At least a dozen photographers" The Man Behind p. 76
86 "Sometimes I think my father" Sports Illustrated, August 21, 1978
86 "All I know is I get what I want" The Man Behind p. 33
87 "The press loves Ted because" The America's Cup p. 45
87 "The reason that I became" Lead, Follow p. 148
87 "I sure as Hell don't" East Bay Window, September 14, 1977
88 "To be amid the scene" The Man Behind p. 33
88 "Somebody put a bottle of aquavit" Tom Snyder Show, October 24, 1977
89 "I never loved sailing" Providence Journal, September 18, 1977
89 "Wouldn't the old man be proud of me tonight?" Citizen p. 217
89 "The New York Yacht Club should have installed" John Bertrand (future America's Cup winner), Born To Win. New York: William Morrow and Co., 1985
89 "Sure I was drunk as a skunk" Lead, Follow p. 12
90 "You ought to catch those Super Bowl" Sports Illustrated, August 21, 1978

NOTES ON SOURCES FROM RESEARCH

90	"I'm overwhelmed by all this" Atlanta Constitution, September 20, 1977
90	"After all, winning the America's Cup isn't as important" Ibid
90	"I had just gone on the air with the Superstation" Lead, Follow p. 150
91	"I think he broke precedent one time by taking us all" Take Me Out p. 192
91	"No matter who wins" Sail, March 1980
92	"It's great to win and it isn't so much fun to lose" Time, August 9, 1982
92	"Ted's strong point" Dennis Conner, No Excuse to Lose. New York: W.W. Norton, 1978
92	"Ted's contribution to Cup awareness" The America's Cup p. 46
93	"Tenacious was alright. It's those dishonest little" John Rousmanierre, Fastnet, Force 10. New York: W.W. Norton and Co., 1980 p. 227
93	"I hope they strapped him below." Citizen p. 244
93	"My worst moment was when" Ibid p. 228
94	"It's no use crying. The king is dead" Sports Illustrated, June 23, 1979
94	"Like any experience" New York Times, August 24, 1979
94	"Was I afraid? I guess I'm more afraid of being afraid" Fastnet, Force 10, p. 228
94	"It was rough, R-U-F-F" New York Times, August 24, 1979
94	"I remember saying to the crew" Fastnet, Force 10, p. 230
95	"I never did enjoy sailing that much." Sports Illustrated, June 23, 1986

CHAPTER SEVEN
TED TURNER DISCOVERS THE WORLD IS ROUND

97	"What's it going to cost?" Citizen p. 160
97	"Sid, I want to buy an uplink." Ibid p. 163
98	"If you were managing one thing in the world" Broadcasting, August 17, 1987
98	"I came up with the concept" Inside p. 23

99	"One of the greatest things I ever wrote" Newsweek, June 16, 1980
99	"My name is R.E. Turner III." Bethesda, Md. Congressional Information Service, 1976
102	"After I debated Gene Jankowsi" Newsweek, June 16, 1980
102	"I want to live five lives." Time, January 6, 1992
102	"It's gonna be tough" Citizen p. 171
102	"Problems are just opportunities waiting to be solved." Television/Radio Age, June 24, 1974
102	"Look what's happened in a year" Esquire, October 10, 1978
102	"The networks are a bunch of pinkos." Newsweek, June 16, 1980
103	"The FCC had to change the rules" Playboy, August 1978
103	"I can do more today in communications" Newsweek, June 16, 1980
103	"Don't be a reckless gambler" Charleston News and Courier, July 17, 1979
103	"Is this enough for you, Dad?" It Ain't p. vii
104	"My father died when I was 24," Time, August 9, 1982
104	"I have four great ambitions." Sports Illustrated, June 23, 1986
104	"I would only run for president if" Broadcasting, April 9, 1990
104	"Someday there will be global" Ibid August 17, 1987
105	"We're all over the world" Ibid
105	"I met with the head of Indian" Ibid
106	"There's still a lot of room to grow" Ibid
106	"The movies that we have" Ibid
106	"We have the biggest and best" Ibid
106	"I'm very ambitious" New York Times, December 16, 1989
107	"People have got to be better informed" Ibid April 9, 1990

CHAPTER EIGHT
CNN CREATES A GLOBAL TALKATHON

109	"It's the only thing that lasts" Time, August 9, 1982
109	"We walk with the ghosts" Lead, Follow p. 17
110	"We don't have to get fat and lazy" Ibid p. 11
110	"In 1978 I started thinking" Inside p. 22
110	"News is really in the dark ages" Ibid p. 30

NOTES ON SOURCES FROM RESEARCH

110	"What would you guys think" Ibid p. 29
110	"Don't you think an all news" Ibid
111	"By then several other services had gotten started" Ibid p. 22
111	"Hi, Ted. I was wondering if" Ibid p. 27
111	"Lachowsky, get a yellow pad" Ibid
112	"Hello, Reese? I'm going to" Ibid p. 2
112	"There are only four things" Citizen p. 228
113	"If something's going on" Inside p. 22
113	"Come on with me. I want to show" Ibid
113	"I knew there was not enough time to do the job" Ibid p. 77
113	"He stirs up so much turmoil" The Man Behind p. 218
114	"PLEASE, TED! DON'T DO THIS TO US!" Inside p. 32
114	"I had met Ted back in April" Ibid p. 75
114	"It was hilarious." Ibid p. 99
115	"He was selling US!" Ibid p. 275
115	"We were all wounded soldiers." Ibid p. 276
116	"We were cocky and arrogant." Ibid p. 168
116	"Reese Schonfeld was the guy" It Ain't p. 192
117	"Nobody thinks we can do it" Lead, Follow, p. 254
117	"I didn't know that satellites" Inside p. 82
118	"T'was three weeks before Christmas" Walter Cronkite, CBS Evening News, December 7, 1979
118	"Carry on" Inside p. 82
119	"RCA had two transponders on Satcom I" Althea Carlson, Dream-O-Vision, The Kid who got Loose in Disneyland. Manuscript in process.
119	"All you guys get out of here" Inside p. 85
119	"Turner ranted, he screamed" Citizen p. 255
119	"Andy, do you own any stock" Ibid p. 253
120	"I remember Ted yelling" Ibid
120	"You're gonna kill me" Ibid
120	"Cangelosi made Ted this offer" Inside p. 86
121	"We will not be stopped!" Ibid
121	"Turner called us all" Ibid p. 92
121	"This is not the solution" Ibid p. 103
122	"Normally, when a television station" Newsweek, February 9, 1987
122	"I said, here's what we'll do" Inside p. 2
123	"We then had a few seasoned" Ibid
123	"This is about it" Lead, Follow p. 257
123	"I'm going to collapse" Atlanta, September 1982

123	"You know, I don't blame" Citizen p. 256
124	"They agreed to stop opposing" Lead, Follow p. 258
124	"I'm going to do news like" Newsweek, June 16, 1980
124	"I'm going to travel" Atlanta Constitution, June 2, 1980
124	"Nobody believed him" It Ain't p. 185
124	"I'd like to call our" Atlanta Constitution, June 2, 1980
126	"Hey, Reese. Is there any place" Lead, Follow p. 242
126	"Much like Muhammed Ali" Newsweek, June 16, 1980
127	"Wait a minute!" Home Video Magazine, June, 1980

CHAPTER NINE
BOY ARE THOSE FOREIGN KINGS AND PRESIDENTS GOING TO BE SURPRISED TO SEE ME!

129	"What I've got to do now" Lead, Follow p. 272
129	"Ted Turner learned more" Atlanta, May 1993
130	"I'm here to serve as the communicator" Time, August 9, 1982
130	"The Global 2000 Report to the President" The Global Report to the President: Entering the Twenty-first Century. Washington, DC: Seven Locks Press, 1980, updated 1988
131	"The report affirmed" Atlanta Business Chronicle, June 12, 1989
131	"If there was a person who educated Ted" Citizen p. 329
131	"Kids today grow up in front of a television set." Paul White Award Acceptance Speech
132	"For the production of television programming" Axel Madsen, Cousteau. New York, Beaufort Book Publishers, 1986 p. 240
133	"We're doing a number of documentaries" Broadcasting, August 17, 1987
133	"When you have critical issues" New Yorker, September 12, 1988
133	"The Society is obtaining funding" Broadcasting, August 17, 1987
133	"It was love at first sight." Richard Munson, Cousteau, the Captain and His World, a Personal Portrait. New York, William Morrow and Co., 1989 p. 109

NOTES ON SOURCES FROM RESEARCH

133	"God bless you." Ibid
134	"How much money do you need?" Ibid p. 190
134	"I gave him $4 million" Playboy, August 1983
134	"Do you think there's hope" Ibid
134	"This man places his money" Cousteau, Madsen p. 233
134	"God bless you" Cousteau, Munson p. 209
135	"I'm a very curious person" Atlanta Constitution, February 20, 1982
135	"I just went down there as citizen" BBC-TV, May 1982
135	"After three drinks" 60 Minutes, April 20, 1986
135	"Castro's not a communist" Playboy, August 1983
135	"I'm the only man on the planet" Sports Illustrated, June 23, 1986
136	"The divers found a sea brimming" Cousteau, Munson p. 214
136	"I am extremely happy to announce" Ibid p. 216
136	"Cousteau saved my life" Ibid
137	"After starting CNN itself" Paul White Award Acceptance Speech
137	"We've finally got them talking" Inside p. 276
137	"As an American" Inside p. 304
137	"Americans never think that there's" It Ain't p. 335
137	"There needs to be peace here." Citizen p. 405
137	"For the next half hour they talked" Ibid
138	"Many of the things" Time, January 2, 1992
138	"Before I could invite them in" Neil Felshman, Gorbachev, Yeltsin and the Last Days of the Soviet Empire. New York: St. Martin's Press, 1992 p. 23
138	"We have a military hot line" Fred Coleman, The Decline and Fall of the Soviet Empire, Forty Years That Shook the World from Stalin to Yeltsin. New York: St. Martin's Press, 1996 p. 25
138	"Boris Yeltsin played the hero." Stuart Loory and Ann Imse, Seven Days That Shook the World, The Collapse of Soviet Communism. Atlanta, GA: Turner Publishing, 1991 p. 230
139	"Citizens of Russia" Ibid
139	"Inside the White House" Ibid p. 231
139	"From this vantage point" Ibid
139	"Day two." Ibid p. 233
140	"There was an elite group" Ibid p. 29
140	"Defenders have locked arms" Ibid p. 233
140	"In the west the story" Gorbachev and Yeltsin p. 20

141 "The plotters of the coup" Ibid p. 31
141 "I wanted to sink into the ground." Ibid p. 32
141 "Gorbachev started by climbing" Gorbachev, Yeltsin p. 263
141 "The motherland of communism" Time, January 2, 1992
142 "Every day we reach more people" Ibid
142 "I mean, it's just good news" Ibid
143 "I don't want to upset the apple cart" Wall Street Journal, August 28, 1986
143 "I think what has occurred has been pretty terrific" Broadcasting, August 17, 1987
143 "I've already met or exceeded" Sports Illustrated, June 23, 1986
143 "I wanted to use communications" Wall Street Journal, February 1, 1990

CHAPTER TEN
SHAKING UP THE NETWORKS

145 "I've got everything I need" Donahue Show, April 1, 1981
146 "It doesn't bother me that I'm committing" Newsweek June 16, 1980
146 "Sure, I'm worried." Time, August 9, 1982
146 "We'll have enough homes" Ibid
146 "My biggest job is in Washington" Atlanta Journal, September, 1982
147 "Cable News Network today files suit" Inside p. 197
147 "I demand an immediate Congressional investigation" Ibid
148 "What those networks are doing" Atlanta Journal, September, 1982
148 "We're going to do a bunch of investigative exposes" Newsweek, June 16, 1980
148 "We're at war with everybody" Atlanta Constitution, May 11, 1981
148 "The worst enemies the United States" Playboy, August 1983
148 "I would first like to point" Inside p. 244
149 "The Big Three are an evil empire." Soundings, July 7, 1977
149 "You know why they're not here any more? Because the mammals" Newsweek, June 16, 1980
149 "There are a lot of things" Broadcasting, August 17, 1987
149 "We've created havoc" Broadcasting, August 17, 1987

NOTES ON SOURCES FROM RESEARCH

150 "You tell me what's going to improve" Ibid
150 "What John Malone said" Ibid
150 "The network ratings this summer are down" Ibid
151 "You take the value of the cable" Ibid
151 "Down in Atlanta, Georgia" Inside p. 251

CHAPTER ELEVEN
THE CABLE NEWS WAR
ABC NEWS AND MGM

153 "Everything I do is a war" Soundings, July 7, 1977
153 "We'd take ninety percent of the receipts due" Contemporary American Business Leaders p. 713
153 "It's a pre-emptive first strike!" Broadcasting, August 19, 1981
154 "It's an immediate counter-attack!" Inside p. 203
154 "Ted's happy. We're at war again." Ibid
154 "It was basically a defense" Ibid p.204
155 "What was your first reaction when you heard that Westinghouse" Ibid p. 205
156 "Holy smoke!" Ibid
156 "When I cast my lot" It Ain't p. 349
157 "Anybody who goes with them" Ibid
157 "I'm really proud of what we've done." Inside p. 207
157 "And now, Ted buddy, it's all yours." Ibid p. 257
158 "I don't know how close he was" Ibid p. 254
158 "How much of it was luck?" Ibid
158 "On a level playing field" Contemporary American Business Leaders p. 713
158 "We might see a little black ink" Lead, Follow p. 272
159 "The office building and retail space" Broadcasting, August 17, 1987
159 "I'll sell you CNN" Citizen p. 336
160 "I'm here to buy you" Ibid p. 337
160 "CBS came down to Atlanta" Ibid
160 "We were invited to come" Ibid
160 "As far as CNN was concerned" Inside p. 265
161 "I needed distribution" Broadcasting, August 17, 1987
161 "Ted's father imbued in him" Citizen p. 338
161 "It's the Tiffany network" Ibid

233

161	"I probably won't make it" Ibid p. 347
161	"It'll be easier for TBS to sell ads" Wall Street Journal, July 26, 1985
161	"We have been very interested in joining forces" New York Times, April 19, 1985
162	"I'll never call them junk bonds" Inside p. 264
162	"I don't know what all this ranting" Atlanta Constitution, May 29, 1985
162	"Poison pill." Wall Street Journal, July 5, 1985
162	"In the end it was worth" Citizen p. 348
162	"It'll be easier for TBS" Wall Street Journal, July 26, 1985
162	"We were pretty solid in news" Broadcasting, August 17, 1987
163	"This is the answer to my dreams." Newsweek, February 9, 1987
163	"To be a long-term factor" Citizen p. 351
163	"Movie fees are rising" New York Times, March 30, 1986
163	"It's a business where the big are getting bigger" Ibid
164	"I paid full value" Ibid
164	"Mr. Kerkorian is no dummy." New York Times, March 30, 1986
164	"I knew we were in trouble before" Broadcasting, August 17, 1987
164	"There were two ways to go" Ibid
165	"I don't think this is the time to be small" Ibid
165	"I owe nearly $2 billion" Wall Street Journal, February 10, 1986
165	"How can you go broke" It Ain't p. 300
165	"Goddamn it! I've really" Sports Illustrated, June 23, 1986
165	"In three minutes Turner had" Newsweek, February 9, 1987
165	"Everybody wants to become a star." Sports Illustrated, August 21, 1978
166	"Dee, a taxi driver in New York recognized" Ibid
166	"Two billion dollars I owe!" Ibid, June 23, 1986
166	"If a major recession caught me" Ibid
167	"NBC gave me until 5 p.m. New York time" Ibid, August 17, 1987
167	"I feel like Scarlett O'Hara" Ibid November 25, 1985
167	"That deal was the road to easy street." Citizen p. 357
167	"Ted is anti-print." Ibid

NOTES ON SOURCES FROM RESEARCH

167   "I don't consider it a rescue at all." Broadcasting, August 17, 1987
168   "Life is a series of gambles" Ibid
168   "They bought stock at 15" Ibid
168   "I've lost control" Citizen p. 381
168   "At the time, the MGM deal" Ibid p. 380
169   "I won't be able to make any large acquisitions" Newsweek, February 9, 1987
169   "Peace on earth, goodwill to men" Citizen p. 446
169   "It's obvious I was seeing things" Wall Street Journal, July 17, 1989
169   "Women put on make-up" Contemporary American Business Leaders, p. 715
169   "Turner's reputation in Hollywood" Newsweek, February 9, 1987
170   "It is as great an impertinence" Ibid
170   "a criminal mutilation" Wall Street Journal, December 24, 1986
170   "cultural butchery" Ibid
170   "There are a lot of the older pictures" Broadcasting, August 17, 1987
170   "I would maintain that the older pictures" Ibid
170   "No need in sitting alone on a shelf" Ibid
171   "All I can say is thank God they shot" New York Times, December 16, 1989
171   "It was a land of cavaliers" Ibid
171   "We're like the Allies" Business Week, July 17, 1989
171   "I have to believe this has worked out" Citizen p. 395
172   "The one thing that we're" Broadcasting, August 17, 1987
172   "I'm looking forward to squishing Rupert" Ted Turner Website, Internet, January 1998
172   "For Ted Turner, the merger" Time, July 29, 1996

CHAPTER TWELVE
TEN PLEDGES TO THE PLANET

173   "I think our major responsibility" Paul White Award Acceptance Speech

CHAPTER THIRTEEN
"SAVE THE EARTH" CAMPAIGN

175   "We're really a pretty terrific species." Citizen p. 319
175   "Turner has been talking about television as a tool" Broadcasting, November 26, 1985
175   "I'm not concerned any longer about myself" Inside p. 272
175   "Like Charlemagne, I'm saving Christendom from the Infidels." Time, August 9, 1982
176   "Population growth is analogous to a plague of locusts." Atlanta Magazine, September 1982
176   "The next ten years are going to be very difficult" Citizen p. 330
176   "It's matter of survival, number one" The Humanist, November, 1989
177   "The President of the United States" Ibid
177   "Ted sees the angles on everything" Citizen p. 421
177   "You bet your bippy" New York Daily News, July 14, 1989
178   "This is the year I was going to get rich" The Racing Edge p. 151
178   "What is a newspaper?" Newsweek, February 9, 1987
178   "I used to hate newspapers" Panorama, April 1980
178   "a newspaper you can watch" Time, August 9, 1982
178   "You know, all you have to do is keep up with what's going on" Inside p. 74
179   "There's peat on that land." It Ain't p. 117
179   "With the increasing numbers of people" Citizen p. 4
179   "Ted was ten years ahead of everybody else" It Ain't p. 407
180   "I think that Turner takes it" Citizen p. 425
180   "I see the whole field of environmentalism" Ted Turner Website, Internet

CHAPTER FOURTEEN
DREAM-O-VISION, TV AT GROUND ZERO

181   "He started thinking about it." Citizen p. 405
181   "Your people are in serious danger" Ibid p. 441
181   "We have a global job to do" Robert Weiner, Live From Baghdad – Gathering News at Ground Zero. New York: Doubleday and Co., 1992 p. 10

NOTES ON SOURCES FROM RESEARCH

181	"As far as I'm concerned, you're all heros." Ibid
182	"You're doing a helluva job" Ibid
182	"If we could somehow pull it off" It Aint p. 370
182	"It's as though we were in training" Live From Baghdad p. 263
182	"CNN's countdown to the deadline" Ibid p. 1
183	"This is it...We're outta here." Ibid
183	"Are you serious? We're using that hotel" Ibid p. 16
183	"The decision to stay in a place" Ibid
184	"At present, we do not have the means" Ibid p. 280
184	"This is Bernie Shaw." Ibid p. 257
185	"You wouldn't believe it." Ibid p. 258
186	"It began in the distance" Ibid p. 283
186	"Of course the Iraqis are the enemy" Wall Street Journal, January 18, 1991
187	"Ted is an internationalist." Inside p. 271
187	"Arnett is probably the toughest" Live From Baghdad p. 17
187	"Both were intensely curious" Ibid p. 290
188	"I want this truck painted like a circus wagon." Ibid p. 292
188	"By the time the Gulf War broke out" Major General Perry M. Smith, How CNN Fought the War, A View From the Inside. A Birch Lane Press Book published by Carol Publishing Group, 1992 p.14
189	"Our best source of intelligence" Wall Street Journal, January 21, 1991
189	"Ted Turner's vision" Forbes, January 4, 1993
190	"For influencing the dynamic" Time, January 6, 1992

CHAPTER FIFTEEN
HOME – AT LAST

191	"I like these ranches." Citizen p. 403
191	"I gotta warn you." Ibid
191	"Buffalo! I want buffalo" Ibid p. 408
191	"Ted's one of the earliest users" Bozeman Daily Chronicle, July 20, 1989
192	"The 'Flying D' is a handsome" Ibid
192	"That's what everyone wants." Montana Magazine, June 1992
192	"I just like watching them" It Ain't p. 385

193	"If they get out" Ibid
193	"This land has unleashed a spiritual" Bozeman Daily Chronicle, November 17, 1991
193	"The only reason those elk are there" Ibid August 6, 1991
193	"I bought this place to get away" Ibid
193	"As far as I can see, it's going great." Larry King Live, October 8, 1980
194	"We have a community of interests." Vanity Fair, April 1991
194	"I find it to be my second most favorite thing to do." Bozeman Daily Chronicle, April 7, 1991
194	"You've got to start with a story" New York Times, September 18, 1993
194	"I love movies." Citizen p. 454
194	"If Ted Turner didn't have Libra rising" The Man Behind p. 42
195	"There wasn't a family in the United States" Citizen p. 455
195	"When I was a little kid – about nine or ten" Ibid p. 27
195	"So, I said, I've got to have a cameo" Ibid p. 456
195	"I'll tell you how I felt." Ibid
196	"Absolutely! I'm telling you!" Ibid
196	"I don't read scripts very much." Ibid
196	"We're going to do Joan of Arc" Ibid
196	"I had MGM" Ibid
197	"Show biz! There's no business" Ibid
197	"I am spending more time in Montana" Montana Magazine, June, 1992
197	"I've invested plenty of bucks" Bozeman Daily Chronicle, March 8, 1992
197	"The 'Flying D' is really becoming" Atlanta Constitution, January 19, 1992
197	"I think Ted is finally convinced he's rich." Citizen p. 403
198	"My father had this burning desire to succeed." Newsweek, June 16, 1980
198	"If I had only one wish" It Ain't p. 207
198	"I want to get my children in the habit of giving" Citizen p. 445
198	"Turner Foundation's" Forbes, October 14, 1996 (Member Forbes 400 List since 1982)
198	"Old Ironsides (USS frigate 'Constitution')" Dream-O-Vision
199	"Its use predates Christopher Columbus" Environmental Magazine, July/August 1996

NOTES ON SOURCES FROM RESEARCH

199	"It's appalling to me" High Times, June 1997
200	"Reducing our demand and use of wood" Ted Turner Website, Internet, January, 1988
200	"You can't get high smoking jeans" US News and World Report, January 20, 1997
200	"Hemp is literally capable of saving the planet" Sierra, November/December, 1995
200	"CNN founder and Time Warner vice chairman" Ted Turner Website, Internet, January, 1988
201	"What I'm trying to do is set a standard" Wilmington Morning Star, Wilmington, NC, September 20, 1997
201	"This is not going to go for administrative" Ted Turner Website, Internet, January, 1988
201	"A billion's a good round number" Ibid
201	"I'm no poorer than I was" Ibid
202	"I'm putting every rich person in the world on notice" Ibid
202	"There are so many rich guys" Ibid
202	"Media mogul Ted Turner" Wilmington Morning Star, Wilmington, NC, September 20, 1997
203	"It's one of my favorite things." New Yorker, September 12, 1988
203	"Turner says that a magazine columnist" Sports Illustrated, August 21, 1978
203	"I want to be the hero" Citizen p. 275
204	"I want to be remembered" Ibid p. 455

CONCLUSION
PRINCE OF THE GLOBAL VILLAGE

205	"Visionaries are possessed creatures." "Prince of the Global Village," Time, January 6, 1992
207	"Build thee more stately mansions, O my soul" Best Loved Poems.

BIBLIOGRAPHY

Alone Together, Elena Bonner. New York: Alfred A. Knopf, 1986

The America's Cup, An Inside View, Bob Bavier. New York: Dodd, Mead and Co., 1986

Bartlett Familiar Quotations, John Bartlett. New York: Little, Brown & Co., 1992

The Berlin Wall, Norman Gelb. New York: Random House, 1986

The Best Loved Poems of the American People, Hazel Felleman. Garden City, NY: Doubleday & Co., 1936

The Birth of Freedom, Shaping Lives and Societies in the New Eastern Europe, Andrew Nagorski. New York: Simon & Schuster, 1993

Brave New World, Aldous Huxley. New York: Time, Inc., 1932

Citizen Turner – the Wild Rise of an American Tycoon, Robert and Gerald Jay Goldberg. New York: Harcourt, Brace and Co., 1995

CNN: The Inside Story – How a Band of Mavericks Changed the Face of Television News, Hank Whittemore. New York: Little, Brown & Co., 1990

Cousteau, Axel Madsen. New York: Beaufort Book Publishers, 1986.

Cousteau, the Captain and his World, a Personal Portrait, Richard Munson. New York: William Morrow and Co., 1989

The Decline and Fall of the Soviet Empire, Forty Years that Shook the World, from Stalin to Yeltsin, Fred Coleman. New York: St. Martin's Press, 1996

Fastnet, Force 10, John Rousmanierre. New York: W.W. Norton and Co., 1980

The Global Report to the President: Entering the Twenty-first Century. Washington, DC: Seven Locks Press, 1980, updated 1988

Gone with the Wind, Margaret Mitchell. New York: MacMillan Publishing Co., 1936

Gorbachev, Yeltsin and the Last Days of the Soviet Empire, Neil Felshman. New York: St. Martin's Press, 1992

The Grand Gesture – Ted Turner, Mariner, and the America's Cup. New York: Little, Brown & Co., 1980

How CNN Fought the War – A View from the Inside, Major General Perry M. Smith. A Birch Lane Press Book published by Carol Publishing Group, 1992

It Ain't As Easy as it Looks, Ted Turner's Amazing Story, Porter Bibb. New York: Crown Publishers, Inc., 1993

Lead, Follow, or Get out of the Way, The Story of Ted Turner. Christian Williams. New York: Times Books, 1981

Life in Russia, Michael Binyon. New York: Pantheon Books, 1993

Wall, Peter Wyden. New York: Simon & Schuster, 1989

Live from Baghdad – Gathering News at Ground Zero, Robert Weiner. New York: Doubleday, 1992

Moscow '86 Goodwill Games, Ken Bastian. Atlanta: The Publishing Group, 1986

No Excuse to Lose, Dennis Conner. New York: W.W. Norton, 1978

On the Brink, the Dramatic Story Behind the Scenes of the Reagan Era and the Men And Women Who Won the Cold War, Jay Wink. New York: Simon & Schuster, 1996

The Racing Edge, Ted Turner and Gary Jobson. New York: Simon and Schuster, 1979

Seven Days that Shook the World – The Collapse of Soviet Communism, Stuart Loory and Ann Imse. Atlanta, GA: Turner Publishing, 1991

1984, George Orwell. New York: Signet, 1950

Television Documentary: Portrait of the Soviet Union, Ira Miskin. Atlanta: Turner Broadcasting, 1988 (seven one-hour segments available from Public Libraries through inter-library loan department)

SELECTED INDEX

A

Aaron, Hank, 64, 68, 123
Abortion for Survival (television show), 177
Academy Awards, 170
Aesop's Fables (classics), 26
Afghanistan, 18
Africa, 129, 130, 142
Air Force One, 135
  Cuba's, 135
  America's, 135
Alexander the Great, 27, 75, 103
Ali, Muhammad (Cassius Clay), 64, 89, 126
Allen, Woody, 170
Allies, 171
Al-Rasheed Hotel, 183, 186
alternative energy sources, 130, 178
Amazon expedition, 134
America's Air Force One, 135
America's Cup, 23, 78-93, 130
  1974 trials, 79
  1977 trials, 81-85
  1977 finals, 85-90
  1980 trials, 91-93
American Broadcasting Company (ABC), 62, 98, 103, 147, 152-158, 183
  and Satellite News Channel, 153-158
  Turner's crusade against, 62, 98, 103, 145-151
  in White House news pool, 147
American Eagle (yacht), 79
Amos, Paul, 186

Ancient Forest: Rage over Trees (television documentary), 177
Angola, 135
Annapolis, (United States Military Academy), 33
Apocalypse Now (movie), 196
Arafat, Yassir, 137
Arbatov, Georgi, 16
Army, Navy, Marine and Air Force bands, 122, 125
Aristotle, 34
archaeological digs, 193
Armageddon, 176
Arnett, Peter, 183-187
Asia, 129, 182
Atlanta Braves, 61-76
  Braves 400 Club speech, 64-65
  promotional gimmicks for, 66, 70
  spring training, 67-68
  Turner's purchase of, 61-62
  in World Series, 63, 74
Atlanta Chiefs, 76
Atlanta Constitution (newspaper), 63, 70
Atlanta Flames, 76
Atlanta Hawks, 75-76
Atlanta LaSalle Corporation, 63
Audubon Society television documentaries, 170
Australia, 63, 130, 165
Australia (yacht), 85-86
Ayatollah, 143
Aye, Calypso (song), 134

B

Babick, George, 126
Baghdad, Iraq, 181-188
Bahamas, 130
Bahouth, Peter, 199-200
Bar None (Montana ranch), 191
baseball, (see Atlanta Braves)
Baseball Commissioner, 72-74, 87
Bat Conservation Society, 198
Baker, James A., 147
Battery Park City, 213
Bavier, Bob, 79, 87, 92
Begathon, 59
Belford, Tom, 177
Bell, Alexander Graham, 203
Ben Hur (movie), 106
Berlin Wall, 22
Bermuda, 130
Bertrand, John, 89
Better World Society, 132-133, 177
Bhopal, 105
Bicentennial, United States of America, 101
Big Daddy, 12
Big Three Networks, (ABC, NBC, CBS), 98, 101-102, 109-110, 120-121, 126, 145-152, 155-163
billboards, 39-50
bison, buffalo, 110, 191-193
Black Pearl, The, 83
Black Sea dracha, 138, 141
Blackfoot Indians, 3, 193
Bogart, Humphrey, 196
Boldin, Valeri, 138
Boston Marathon, 143
boycotts, 20
Boys and Girls Clubs of America, 23
"Boys from Baghdad, The," 185
Bozeman, Montana, 191-197
Bozeman cattle ranchers, 192-193
Boznia-Herzegovina, 195

brass bands, 90, 122, 125
  Army, Navy, Marine and Air Force, 122, 125
  hero's welcome in Atlanta, 90
Brazil, 130
Brer Rabbit, 168, 211
British Broadcasting Company (BBC), 103, 105, 133
Broadcasting Magazine, 175
broadsword, 2, 102, 121
Brown, Lester R., 17
Brown University, 32-37
Buckley, William F., Jr., 16
buffalo (bison), 110, 191-193
Bush, George, 137
Business Alliance for Commerce in Hemp, 200

C

Cable News Network (CNN), 2, 7-8, 10, 16-17, 25, 109-142, 181-190
  creation of, 109-127
  CBS takeover attempt, 159-162
  and new CNN Center in Omni complex, 159
  and CNN launch, 124-127
  and White House news pool, 147
  Gorbachev attempted coup coverage by, 138-142
  Headline News Service, 154
  international operations of, 2, 7-8, 10, 16-17, 25, 109-142, 181-190
  Larry King Live interviews, 193, 201
  NBC attempts to buy, 167
  Persian Gulf War coverage by, 181-190
  TBS board of directors, 167-169
  World News Report (television show), 105-106, 137
  USA Today attempts to buy, 167

SELECTED INDEX

cable industry, 97-107, 111, 156
Paul White Award for Television Excellence, 131
cable television, 150-151
descramblers, 150
wiring the country, 150
wiring the globe, 104
wiring New York and Washington, DC, 150
Cambodia, 142
Capitalism, 19, 36, 211
Capra, Frank, 146
Captain Cook, 27
Calypso (research vessel), 134, 136
Cangelosi, Carl, 120
Canadian Imperial Bank, 159
Carlos, Juan, Prince of Spain, 51
Carousel (movie), 86
Carter, Jimmy, 85, 206
Castle Rock Entertainment, 196
Castro, Fidel, 19, 135-136
Catholics, 19
cattle ranchers, 192
Central Intelligence Agency (CIA), 136
Chase Manhattan Bank, 166
Chambered Nautilus, The (Oliver Wendell Holmes), 207
Chaney, Darrel, 91
Channel 17 (UHF station), *(WJRJ) (WTCG) (WTBS) (Superstation)*, 98-107, 109-111, 134, 150, 159, 161, 163, 172
chemical and biological weapons, 132
Chief Nok-A-Homa, 66
Chinese television, 133
China Syndrome (movie), 196
Choate (prep school), 33
Christendom, 175

churches
of Newport, 88
Roman and Greek, 34
Civil War, 79, 195
Classics, The, 33
Clay, Cassius (Muhammad Ali), 64, 89, 126
Cold War, 2
colorization, 169-170
Columbia Broadcasting System (CBS), 98, 102-103, 105, 118, 147, 152, 159-162
Turner crusade against, 62, 98, 103, 145-151
takeover bid for, 159-162
in White House news pool, 147
Columbus, Christopher, 1, 36, 136, 169, 199
Communists, 19
Concepcion, Dave, 64
confederate saber, (broadsword) 2, 102, 121
Congress, U.S., 2, 102, 121
Congressional hearings, 102
Congressional investigation, 147
Conner, Dennis, 89, 92
conservation easements, 191-192
conservationist, 192
Constantine, King of Greece, 51
Constitution (frigate), 25, 27, 198
Courageous (yacht), 81-88
Cousteau, Jacques Yves, 133-134, 136
Cousteau Society, 134
Cowes, England, 93
Coxe, Tench, 120
Cracker Jacks, 65
Craig, Marilyn, 200
Cronkite, Walter, 118, 162, 183
Crossfire (television show, CNN), 16
Cuba, 78, 130, 135-136
Cuba's Air Force One, 135

Cuomo, Mario, 23
Curle, Chris, 155

D
Daily Worker, 16
Dames, Peter, 32
Dark Ages, 110
Dark Shadows (movie), 31
Delta, rocket, 118
Denmark, 51, 130
Denver, John, 133-134
Dinner With Andre (movie), 1
dinosaurs, 3, 149
Dirty Dozen, 116
disarmament, 16, 18, 174, 176
Disneyworld, 75
Dodgers, 126
Don Quixote, 15
Donahue, Dan, 61
Doyle, Larry, 183
dracha, Black Sea, 138, 141
dreadnoughts, 27
Drug Enforcement Administration (DEA), 199

E
ESPN, 112
Eason, Tim, 185
Ebeneezer Scrooge Prize, 203
economics, 36
Edison, Thomas, 192, 203
Egypt, 130, 137
Einstein, Albert, 62
elk, trophy, 3, 193
Elliot, Sam, 196
energy sources, alternative, 130, 178
England, 93-95
English Channel, 95
Enterprise (yacht), 81-84
Europe, 106, 129, 142-143, 158, 182
explorer, 2, 31, 203, 207

F
Farmer, Don, 155
Fastnet Race, *(Fastnet of Death),* ocean sailing race, 93-95
Faulkner, William, 35
Federal Communications Commission (FCC), 99, 102
First National Bank of Chicago, 45
Flying D (Montana ranch), 191-197
Flynn, Errol, 79
Fonda, Jane, 3, 193-197
Forbes Magazine, 198
   Forbes 100 List, 202
Ford, Henry, 192
Forester, C.S., 26
Fortune 500 Club, 18
Frankfurt, 105
freedom of speech, 147
Friends of the Wild Swan, 198
Fulton Stadium, 61

G
game wardens, Indian, 192
Garden of Eden, 110
Gearon, Mike, 158, 162, 177
Georgetown University, 103
Georgia Military Academy, 28
Gettysburg (battle), 195
Gettysburg (movie), 195
Giants, San Fransisco, 71, 73
Gilligan's Island (television show), 101
Global Leadership Award (United Nations), 201
Global 2000 Report to the President (Congressional study), 130-131
Gone with the Wind (movie), 165, 171
Goodwill Games, 1-2, 8-21
   Goodwill Games, Moscow, 1-2, 8-21
   Goodwill Games, New York, 23 appendix, 213-215

SELECTED INDEX

Goodwill Games, Seattle, 21-22
Goodwill Games, St. Petersburg, 22-23
Goodwill Games Aquatic Center, 214
Gorbachev, Mikhail, 1-2, 13, 17-18, 137-142
Gorbachev, Raisa, 138
Goring, Hermann, 162
Gorman, Pat, 123
Gostelradio (Russian radio and television), 15
Grand Europe, Hotel (St. Petersburg), 22
Greased Lightning (sailboat), 77
Great Wall of China, 137
Greece, 51, 130
Greenpeace, 199
Greek literature, 26, 34-36
Gulf War, 181-190
Greek and Roman mythology, 26

H

Hannibal, 99
Hart, Lois, 116
Headline News, (TBS) 154
Heart of Gold Award, 203
Helfrich, Bunky, 93, 113
Hemp, 198-200
Hemp Museum, 198
hero, 2, 65, 90, 103, 145, 203, 207
Hesburgh, Ted, 21
Hinkley, John, Jr., 148-149
Hinman, George, 89
Hitler, Adolph, 142, 148, 167
Hogan, Gerry, 55, 102, 115
Hollywood, 149, 163-165
Holliman, John, 185
Holmes, Oliver Wendell, Justice, 25, 207

Home Box Office (HBO), 111-112, 120
Home Video Magazine, 127
Hope, Bob, 73
Hope Plantation, 109, 129
Hotel Al-Rasheed, 183, 186
Hotel Grand Europe (St. Petersburg), 22
human species, survival of, 131, 133, 173, 175, 180
Hurst, Steven, 138-140
Huston, John, 170
Hussein, King of Jordan, 137, 187
Hussein, Saddam, 186

I

I Love Lucy (television show), 101
I Was Cable When Cable Wasn't Cool (promotional song), 151-152
India, 36, 105, 130
Indian game wardens, 192
Indians, Blackfoot, 3, 193
infidels, 175
Ingles, Andrew, 119
Intrepid, U.S.S., aircraft carrier, 214
Iran, 18, 105
Iraq, 181-190
Iron Curtain, 2
Israel, 130

J

Jankowski, Gene, 102
Japan, 130
Japanese, 148
Jews, 12, 16, 19
Jiminy Cricket, 104, 203
Joan of Arc (movie), 196
Jobson, Gary, 81
Jordan, 130, 186
Justice Department, 119

K

KGB agents, 2, 19, 138, 140-141
Kavanau, Ted, 116, 117, 121-122, 154
Kennedy, John, 74
Kerkorian, Kirk, 164
Knots Landing (television show), 106
Korea, 9, 20
Kremlin, The, 2, 12-13
Kuhn, Bowie, 72-74, 87

L

Lachowski, Don, 111
Lacota Woman (television movie), 196
Lake Erie, 77
Lake Lanier, 23
Larry King Live (television show), 193, 201
Lawrenceville (prep school), 33
Leach, Kathy, 137
Lebanon, 18
Lee, Robert E., General, 28
Leningrad (St. Petersburg), 22-23
Leonidas, 35
Li-Peng, Premier of China, 137
Libra (astrological sign), 194
Lincoln, Abraham, 74
London, England, 105
Love Boat (television show), 110
Lucas, Bill, 69

M

MCA, 59
Macedonians, 27
Madison Avenue, 161
Madison Square Garden, 213
Mafia, 64, 110
Malone, John, 150
Man of the Year, 1991, *Time* magazine, 190, 205
Manigault, Peter, 191
Manilus (military prep school), 32
Mariner (yacht), 79-80
Marines at Tripoli, 27
Matthews, Gary, 71-72
May Day, (military parade in Moscow), 12
Mazo, Irwin, 42-43, 45, 47, 53
McCallie School (military prep school), 26-31
McClurklin, Lee, 53
McGinnis, Richard, 48
McGraw, Tug, 70
McGuirk, Terry, 67, 117, 120, 189
McLuhan, Marshall, 205
Men Against the Sea (Nordoff and Hall), 26
Messersmith, Andy, 70
Metro Goldwyn Mayer (MGM), 163-171
Metropolitan Museum of Art, 162
Mexico, 12, 130
Meyrowitz, Joshua, 138
Miami Vice (television show), 106
Michaelangelo, 49
Middle East, 137, 181-190
Mikanos, 130
military hot line, 138
Minnesota Twins, 74
Miskin, Ira, 15
missiles, 1, 22
missionary, 30
Mitchell Athletic Complex, 213
Molokai, Hawaii, 78
molotov cocktails, 139
Montana, 191-197
Montego Bay race, 78
Moscow, 1-2, 7-15, 19-23, 105, 138-142
Moscow University, 18
Motion Picture Association of America, 103
Mount Vernon, 134

SELECTED INDEX

Murdoch, Rupert, 167, 172
Mutiny on the Bounty (Nordoff and Hall), 26
mythology, Greek and Roman, 26

N
NASA, 117-118
National Basketball Association (NBA), 103
NSA, 189
National Anthem, 121-122, 125
Naegele, Bob, 43, 53
Naegele Companies, 46
Napoleon, 12, 27, 178
Nassau, Bahamas, 117, 130
Nassau Veterans Memorial Coliseum, 214
National Basketball Association (NBA), 103
National Broadcasting Company (NBC), 98, 103, 119-123, 152, 167
  Turner crusade against, 62, 98, 102-103, 145-151
  in White House news pool, 147
National Geographic, 161, 170
Nazis, 148
Nearer My God to Thee (hymn), 121-122, 206
New Line Cinema (movie studio), 196
New York City, 23, 54, 100, 105, 126, 150, 160, 164, 166, 179, 183, 200
New York Yacht Club, 79-80, 87, 89, 91, 164
New World, 1, 27, 106, 111
Newlywed Game, The (television show), 110
Newport, Rhode Island, 79-89
Newsweek magazine, 126, 169
Nicaragua, 135
Nigger of the Narcissus (poem), 79
Noah, 95

Nordhoff and Hall, 26
Normandy, France, 171
Northern Ireland, 18
nose ball race, 70
Noyes, Gale, 36
nuclear briefcase, 138
nuclear war, 17
Nur, Queen of Jordan, 187

O
ocean racing, 78-95
O'Hara, Maureen, 58
Olaf, King of Norway, 51
Old Ironsides (frigate *USS Constitution*), 25, 27, 198
Olympic Games, 9, 10, 20
  Olympics, Los Angeles, 9
  Olympics, Montreal, 10
  Olympics, Seoul, Korea, 9, 20
On Golden Pond (movie), 196
Omni complex (new CNN headquarters), 159
Oscars, 170
ostrich race, 70
Outdoor Advertising Association of America (OAAA), 46-47

P
Paley, William, 162
Paramount Pictures, 59
Patterson, Houston, 29
Paul White Award for TV Excellence, 131
peace, 1, 2, 7-8, 11, 14, 18-19, 125, 133, 137, 142, 169, 177, 181-182, 201, 206
peat bogs, 178-179
Penguin (class of sailboats), 207
People Magazine, 112
Pepsi-Cola, 21
Persian Gulf War, 181-190

Philadelphia '76ers, 75
Philistines, 36, 169
Pickett's Charge, 195
Pike, Sid, 60
Pitcairn's Island (Nordhoff and Hall), 26
Plato, 34
pollution, 130, 200
Ponzi scheme, 48
population control, 104, 107, 130, 132-133, 137, 142, 176
Portrait of the Soviet Union (television documentary series), 1, 15-16
President of the United States, on running for, 104
Programming, Channel 17, 59
Public Broadcasting System (PBS), 133
Pulitzer Prize, 187
Pyle, Barbara, 130

R
RCA, 118-121
racing, 60, 78-95, 130
   America's Cup, 78-93
   at Brown University, 33, 77-78
   Denmark, 51
   Fastnet Race, 93-95
   on Lake Lanier, 23
   Lightning class, 77
   at McCallie School, 78
   Montego Bay, 78
   Nassau, 117
   ocean, 78-95
   Penguins, 77
   SORC, 60, 78-79
   Sydney-Hobart, 79
Rather, Dan, 113
Reagan, Ronald, 1, 13, 137, 147-148, 155
Reasoner, Harry, 68

Red Square, Moscow, 12
Rio de Janeiro, 12, 179
Robinson, Eddie, 91
Robinson Humphrey brokerage house, 53
Rockefeller Center, 119
Rocky (theme song from movie), 83, 90
Roddy, Jim, 56
Roosevelt, Franklin D., 154
Roman and Greek churches, 34
Rome, 99
Rose, Pete, 64
Russian Revolution, 2
Russian television, 15

S
saber, confederate, 2, 102, 121
Safer, Morley, 187
San Francisco Giants, 71
Sao Paolo, 12
Satcom (satellites), 117-121
Satellite News Channel (SNC), ABC, 153-158
satellite transponders, 97-121
Savannah Beach, SC, 25
"Save the Earth" campaign, 173-186
Scarlett O'Hara, 167, 196
Schonfeld, Reese, 60, 111-126, 154
   and creation of CNN, 112-126
   first meeting with Turner, 60
   and Headline News, 154
   and satellite transponders, 97
   and suit against White House, 147
   at war with RCA, 118-121
Schussler, Bob, 58
Schwab, David, 193
Scorpio (astrological sign), 194
Scotland, 130
Senate, Subcommittee on Communications, 99-102

SELECTED INDEX

Seoul, Korea, 9
Seriphos, 130
Shales, Tom, 16
Shaw, Bernard, 181-190
Sheen, Martin, 196
Shephard, Jim, 116
Shipman, Claire, 139
Showtime, 111
Sieber, Bob, 102
Sistine ceiling, 49
Smith, Perry M., Major General, 188
soil erosion, 130
South Africa, 142
South Korea, 20
Southern Ocean Racing Circuit (SORC), 60, 78-79
Soviet Union, 1-2, 7-23, 138-142
Soviet satellite, 139
Soviets, 1-2, 7-23, 138-142
Spanish Armada, 95
Speakes, Larry, 147
spectrum fee, 149
Sports Illustrated magazine, 112
Star Spangled Banner, 121-122, 125
State of the World (annual report of the Worldwatch Institute), 17
Stewart, Jimmy, 170
Strong, Maurice, 179
success, 53, 103-104, 143, 198
Success Magazine, 103
Super Bowl, 90
Superpowers, 1
Superstation (Channel 17) (WJRJ) (WTCG) (WTBS) (UHF station), 98-107, 109-111, 134, 150, 159, 161, 163, 172
survival, 79, 109, 132, 176-177, 180
 of the planet, 132, 176
 of the species, 176-177, 180
 of the fittest, 109

Sydney-Hobart race, 79
Systemic Lupus Erythematosus, 31

T

Tahiti, 63, 130
Take Two (television show), 155-156
Tasmania, 78
Taxi Driver (movie), 101, 148
Tenacious (yacht), 69, 82, 93-95
Tennessee State debating champion, 30
Thatcher, Margaret, 137
Third World, 18, 165
Tianenmen Square, 140
Tiffany network (CBS), 161
Titanic (ship), 107
Time magazine, 84, 111, 190, 205
 Time magazine Man of the Year, 1991, 190, 205
Time Warner, 172
Today Show, The, (television show), 90
Tom and Jerry cartoons, 106
Tomahawk cruise missile, 186
Toyota, 20
Trahey, Jim, 102
Trident submarine, 107
Turner, Ed, (no relation), 120, 182
Turner Family Foundation, 198-199
Turner, Florence Rooney, (mother), 77, 166
Turner, Fonda, Jane, 3, 193-197
Turner, Nye, Judy, 39-40, 46, 48
Turner, Robert Edward, Jr. (Ed) (father), 2, 26, 33-36, 39, 40-46, 60, 62, 72, 80, 86, 89-90, 104, 161, 178, 195, 198
 as enduring presence, 42, 44, 46, 60, 62, 72, 80, 86, 89-90, 104, 161, 178, 195, 198
 suicide of, 41-42

251

as teacher and disciplinarian, 2, 26, 33-36, 39-40
Turner, Robert Edward, III, (Ted), 1-211
  and Atlanta sports teams, 61-76, 81-82, 85, 90-91, 126, 153, 197
  and *Bar None* ranch, 191
  broadening of, 129-143
  at Brown University, 32-37
  CBS takeover bid of, 159-162
  childhood of, 25-37
  CNN launched by, 109-127
  colorization of movies by, 169-170
  in Cuba, 135-136
  and death of sister, 31-32
  editorializing by, 145-153, 175-180
  environmental concerns of, 175-180
  and father's suicide, 41-44
  and *Flying D* ranch, 191-197
  at Georgetown University, 103
  at Georgia Military Academy, 28
  at *Gone with the Wind* gala, 171
  and Goodwill Games, 1-2, 8-21
  and Persian Gulf War, 181-190
  at McCallie School, 26-31
  MGM acquired by, 163-171
  and Mikhail Gorbachev, 1-2, 13, 17-18, 137-142
  as movie producer, 194-197
  and outdoor advertising business, 39-50
  and reorganization of TBS, 167-169
  Turner Pictures, 196
  and sailing, 60, 78-95, 130
  and Satellite News Channel (ABC, Westinghouse), 153-158
  testifies before Congress, 99-102
  UHF television stations purchased by, 53-59
  at White House dinner, 85
  and White House news pool, 147

Turner, Smith, Jane, 51-52, 59, 85, 118
Turner, Teddy, 197
Turner Advertising Companies, 46
Turner Broadcasting Systems (TBS), 105, 134, 150, 159-172
  CBS takeover bid of, 159-163
  environmental issues, 18-20, 130-134, 173-186
  Goodwill Games, 1-2, 18-21
  headquarters, 105, 114, 159
  MGM takeover by, 163-166
  satellites, 97-121
Turner Pictures, 196

U

UHF television, 22, 53-61, 99-102
U.S.S.R., 1-2, 7-23, 138-142
UNICEF, 23, 201
US News and World Report magazine, 200
USA Today newspaper, 167
USS Constitution (frigate), 25, 27, 198
USS Intrepid, aircraft carrier, 214
United Artists, 59
United Nations, 8, 21, 125, 174, 179, 200-203
  Earth Summit (Rio de Janeiro, 1992), 179
  Iranian ambassador to, 105
  Turner billion dollar donation to, 8, 200-203
  UNICEF, 23, 201
United States Naval Academy (Annapolis), 33
U.S./Soviet relations, 1-2, 7-23, 138-142
Urals, 16

V

Vamp-X (yacht), 78
Vaughan, Roger, 85, 88
Verdun, 12

SELECTED INDEX

Veterans of Foreign Wars, 148
Viacom, 59
Vietnam, 18, 142, 187

W

WRET (Channel 36), 58-59, 99, 117, 123-124
WTBS (Channel 17), see WTCG
WTCG (Channel 17) (WTBS) (Superstation), 98-107, 109-111, 134, 150, 159, 161, 163, 172
 environmental programming on, 104-107, 130-134
 FCC restrictions and licensing, 99, 102
 purchase of, 53-57
 sports on, 61-76, 81-82
War of 1812, 27
Washington, George, 27, 198
Washington, D.C., 110, 146, 150, 155, 160, 183
Washington Post newspaper, 16
Waterloo, 12
weapons, 13, 16-18, 132, 176
 chemical and biological, 132
 nuclear, 13, 16-18, 132, 176
Wein Stadium, 214
Weiner, Robert, 182-188
Westinghouse, 117, 123, 148, 153, 155, 158
white horse, 3, 5, 7, 148

White House, Russian, 138-139
White House, US, 1, 66, 85, 147-148, 181, 183
white (haunted) house, 114-117
Williams, Mary Alice, 114
Williams, Pat, 75
Williams, R.T., 54
Wizard of Oz, The, (movie), 106
Woods, Dee, 114
World News Report (television show), 105-106, 137
World War I, 27
World War II, 27, 148, 158
World Series, 63, 74
Worldwatch Institute, 17
Wounded Knee, (battle), 196
Wright, Gene, 56
Wowuka, 3
Wussler, Robert, 9-11, 13, 129, 161, 167, 187
 and Goodwill Games, 9-11, 13
Wyman, Thomas, 160

Y

Yazov, Russian Deputy Defense Minister, 141
Yeltsin, Boris, 138-141
Youth (poem), 79
Yushkiavitshus, Henrikas, 11, 15

Z

Zook, John, (Falcons), 64

REQUEST FOR INFORMATION ON TED TURNER

For a manuscript in process, "Dream-O-Vision – The Kid Who Got Loose in Disneyland," the author requests any first hand knowledge of Ted Turner as a benefactor. It has become apparent that he gives financial assistance without publicity to many people in need and particularly to many people with a grass roots crusade.

If you have a story to tell, please write the author c/o the publisher.

REQUEST FOR ATHLETES WHO LOST THEIR CHANCE AT A MEDAL DUE TO US/SOVIET OLYMPIC BOYCOTTS

"Dream-O-Vision" sets the life of Ted Turner against the backdrop of the Cold War, beginning with the Berlin Wall, the coup d'etat on Nikita Khruschev before the days of CNN when all Soviet information was under total control of the Soviet government, and includes the story of Andrei Sakharov, Russia's top atomic and hydrogen bomb scientist. His arrest by the KGB when he became Russia's leading peace activist and deplored the invasion of Afghanistan prompted President Jimmy Carter's boycott of the 1980 Moscow Olympics and the subsequent Soviet boycott of the 1984 Los Angeles Olympics. The KGB harassment of The Nobel Peace Prize winner Sakharov while under house arrest in Gorky and his hunger strike protests is a poignant story, and will be fully covered in this book.

The effect of the boycotts on individual athletes who were in their prime during the height of the Cold War is equally poignant. There was a ten year hiatus between the 1978 Moscow Olympic Games and the 1986 Goodwill Games when 54% of the top medal contenders could not compete with each other, a disastrous gap in the life of an athlete. The stories of individual athletes who were not given their chance to compete after a lifetime of training would be greatly appreciated.

If you have a story to tell, please write the author c/o the publisher.

Thank you.

Althea Carlson   c/o Episcopal Press
222 Seacrest Drive
Wrightsville Beach
NC 28480
910-256-3537

EPISCOPAL PRESS

If you liked this book, please tell a friend.
Postcard enclosed.

RIDING A WHITE HORSE

ORDER FORM

We wait for all checks to clear before shipping. This includes Priority Mail orders. If you want to speed delivery time, please send a U.S. money order or use Master Card or Visa. Those orders will be shipped right away.

Complete this order form and send with payment or credit card information to:

EPISCOPAL PRESS
222 Seacrest Drive
Wrightsville Beach, NC 28480

Name

Address

City and State

Zip code

Daytime Phone (in case we have a question)   (   )

_____ This is my first order   _____ I have ordered before

Method of Payment _____ Money Order _____ Check _____ Visa
_____ Mastercard

#_____   Expiration Date _____

Title: *RIDING A WHITE HORSE – TED TURNER'S GOODWILL GAMES AND OTHER CRUSADES*, Althea Carlson

Quantity _____ Price $19.95 Subtotal	_____
For delivery in NC add 6% tax ($1.20 per book)	_____
Shipping and handling add $3.00 per book	_____
For Priority Mail add an additional $2.00 per book	_____
Total	_____

Thank you for your order. We appreciate your business.

ABOUT THE AUTHOR

Althea Carlson was born in New York City in 1938, the same year as Ted Turner. She was Executive Director of the World Trade Center Wilmington, interacting with 156 World Trade Centers via the World Trade Center Association on-line computer NETWORK and at worldwide WTCA conventions. She is a business writer, editorial consultant for books and a local newspaper.

She attended Greenwich Country Day School and received a classical education at Bishop Strachan School, Toronto, Ontario, Canada. She received a degree in Mathematics, Statistics and Economics from the University of Connecticut, and did graduate work at New York University.

Her hobbies are sailing with her daughter in a 41 foot Morgan on Long Island Sound, and boating with her sons near her home on the Intracoastal Waterway in Wrightsville Beach, North Carolina.